MW01486845

Higher Oil Prices
and the World Economy

EDWARD R. FRIED
CHARLES L. SCHULTZE
Editors

Higher Oil Prices and the World Economy
The Adjustment Problem

EDWARD R. FRIED *and* CHARLES L. SCHULTZE

GEORGE L. PERRY

GIORGIO BASEVI

TSUNEHIKO WATANABE

WOUTER TIMS

JOHN WILLIAMSON

JOSEPH A. YAGER *and* ELEANOR B. STEINBERG

THE BROOKINGS INSTITUTION
Washington, D.C.

Copyright © 1975 by

THE BROOKINGS INSTITUTION

1775 Massachusetts Avenue, N.W., Washington, D.C. 20036

Library of Congress Cataloging in Publication Data:

Main entry under title:

Higher oil prices and the world economy.

   Includes bibliographical references and index.
   1. Petroleum products—Prices—Addresses, essays,
lectures.   2. Economic history—1945– —Addresses,
essays, lectures.   3. International economic
relations—Addresses, essays, lectures.   4. Petroleum
industry and trade—Addresses, essays, lectures.
I. Fried, Edward R.   II. Schultze, Charles L.
III. Brookings Institution, Washington, D.C.
HD9560.4.H53   338.2′3   75-34234
ISBN 0-8157-2932-4
ISBN 0-8157-2931-6 pbk.

1 2 3 4 5 6 7 8 9

Board of Trustees

Douglas Dillon
Chairman

Louis W. Cabot
Chairman, Executive Committee

Vincent M. Barnett, Jr.
Lucy Wilson Benson
Edward W. Carter
George M. Elsey
John Fischer
Kermit Gordon
Huntington Harris
Roger W. Heyns
Luther G. Holbrook
William McC. Martin, Jr.
Robert S. McNamara
Arjay Miller
Barbara W. Newell
Herbert P. Patterson
J. Woodward Redmond
H. Chapman Rose
Warren M. Shapleigh
Gerard C. Smith
Phyllis A. Wallace
J. Harvie Wilkinson, Jr.

Honorary Trustees

Arthur Stanton Adams
Eugene R. Black
Robert D. Calkins
Colgate W. Darden, Jr.
Marion B. Folsom
John E. Lockwood
John Lee Pratt
Robert Brookings Smith
Sydney Stein, Jr.

THE BROOKINGS INSTITUTION is an independent organization devoted to nonpartisan research, education, and publication in economics, government, foreign policy, and the social sciences generally. Its principal purposes are to aid in the development of sound public policies and to promote public understanding of issues of national importance.

The Institution was founded on December 8, 1927, to merge the activities of the Institute for Government Research, founded in 1916, the Institute of Economics, founded in 1922, and the Robert Brookings Graduate School of Economics and Government, founded in 1924.

The Board of Trustees is responsible for the general administration of the Institution, while the immediate direction of the policies, program, and staff is vested in the President, assisted by an advisory committee of the officers and staff. The by-laws of the Institution state: "It is the function of the Trustees to make possible the conduct of scientific research, and publication, under the most favorable conditions, and to safeguard the independence of the research staff in the pursuit of their studies and in the publication of the results of such studies. It is not a part of their function to determine, control, or influence the conduct of particular investigations or the conclusions reached."

The President bears final responsibility for the decision to publish a manuscript as a Brookings book. In reaching his judgment on the competence, accuracy, and objectivity of each study, the President is advised by the director of the appropriate research program and weighs the views of a panel of expert outside readers who report to him in confidence on the quality of the work. Publication of a work signifies that it is deemed a competent treatment worthy of public consideration but does not imply endorsement of conclusions or recommendations.

The Institution maintains its position of neutrality on issues of public policy in order to safeguard the intellectual freedom of the staff. Hence interpretations or conclusions in Brookings publications should be understood to be solely those of the authors and should not be attributed to the Institution, to its trustees, officers, or other staff members, or to the organizations that support its research.

Foreword

THE SERIES of sharp increases in the price of oil that began in October 1973 subjected the world economy to a severe shock that proved difficult to manage not only because of its suddenness and size but also because it was imperfectly understood. Its repercussions continue to affect the world economy as adjustment policies—both domestic and international—continue to be debated.

How great has the economic cost of higher oil prices been and what forms does it take? How well grounded is the widespread fear of intolerable financial strain that it has engendered? How great is its threat to living standards and economic welfare generally? What domestic and international policies would enable the importing countries to reduce the damage to themselves and to the world economy as a whole?

To address questions such as these, the Brookings Institution commissioned a series of papers examining the effects of the quadrupling of oil prices on the United States, Western Europe, Japan, the developing countries, the international financial system, and the international oil market. The papers were discussed at a conference held at Brookings in November 1974. They were later revised to reflect the conference discussion and, wherever feasible, subsequent developments. The papers, together with an introductory chapter that sets forth the analytical framework of the study, summarizes its conclusions, and offers policy prescriptions, form the contents of this volume.

Three principal conclusions emerge from the study. First, the quadrupling of oil prices was a major cause of the economic recession from which the industrial countries are only now beginning to emerge, and continues to be a serious restraint on economic development in the developing countries that do not produce oil. These are held to constitute by far its most damaging consequences. Second, the fear that higher oil prices will create inherently unmanageable problems for the international financial system or seriously threaten living standards and economic growth in the industrial countries is exaggerated. Third, the main tasks of adjustment policy are the proper management of aggregate demand in the industrial countries and the provision of special financial assistance to the developing countries.

The conference was managed and the introductory chapter was written by the editors, Edward R. Fried, a senior fellow in the Brookings Foreign Policy Studies program, and Charles L. Schultze, a senior fellow in the Brookings Economic Studies program and professor of economics at the University of Maryland. The conference participants principally responsible for commenting on the topics with which the remaining chapters deal are as follows: Saul Hymans of the University of Michigan (the United States), Pieter de Wolff of the University of Amsterdam (Western Europe), Hisao Kanamori of the Japan Economic Research Center and Masahiro Sakamoto of the National Institute for Research Advancement (Japan), Jagdish N. Bhagwati of the Massachusetts Institute of Technology and Carlos Díaz-Alejandro of Yale University (the developing countries), Andrew D. Crockett of the International Monetary Fund and Rimmer de Vries of the Morgan Guaranty Trust Company (the international financial system), and Hendrik S. Houthakker of Harvard University and James T. Jensen, a consultant (the international oil market).

This study was supported by a grant from the Ford Foundation. The editors thank Fred Sanderson and Frank W. Schiff for reviewing and commenting on the manuscript. James Becker verified its factual content and provided research assistance. David Howell Jones edited the manuscript; Julia McGraw prepared the index. Brenda Dixon and Janet Smith provided administrative support for the conference and typed the manuscript in early drafts and in its final version.

The views presented in this book are those of the authors and should not be ascribed to the Ford Foundation or to the trustees, officers, or other staff members of the Brookings Institution.

KERMIT GORDON
President

October 1975
Washington, D.C.

Contents

Tables

Figures

Higher Oil Prices
and the World Economy

CHAPTER ONE

Overview

EDWARD R. FRIED *and* CHARLES L. SCHULTZE
Brookings Institution

NO EVENT OF THE PERIOD following the Second World War had so sharp and pervasive an impact on the world economy as the series of shocks to the oil market that followed closely on the outbreak of the Arab-Israeli war on 6 October 1973. For more than a year, uncertainty grew and pessimism seemed to feed on itself. Indeed, the economic, political, and psychological repercussions of what became known as the oil crisis caused widespread questioning of the capacity of the world economy to adjust to the new situation at tolerable cost.

At first the threat of a deepening supply shortage was the major element of concern. Arab exporters supplied more than 40 percent of all the oil consumed in the non-Communist world. Consequently, when the Arab countries announced, first, that they would reduce their exports by 10 percent from the September 1973 level until their political demands were met, and subsequently, that the cut would be increased to 25 percent, the oil-importing countries faced the possibility of substantial economic dislocation lasting for an indefinite period. The oil shortage, however, turned out to be less severe than had been anticipated. Not all the Arab countries reduced exports according to plan, some non-Arab countries were able to increase production, and most important, the restrictions were relaxed fairly soon. For the period from October 1973 to March 1974 as a whole, the reduction in total oil exports was equal to perhaps 7 percent of consumption requirements in non-Communist countries. By April 1974, for all practical purposes the use of the oil weapon had ended.

Thus, doubts about the economic future came to rest heavily on the

1

world's capacity to live with higher oil prices. Within three months of the outbreak of the war the export price of the standard quality oil (Saudi Arabian light crude oil), calculated in constant dollars, was 3½ times what it had been before October 1973. Subsequent adjustments during 1974 brought the price to approximately four times the precrisis level. This increase in the price of oil meant that in 1974 additional export receipts of the oil-exporting countries were close to $75 billion, an amount equal to approximately 2 percent of the national output of the importing countries. So large and sudden a transfer of income posed difficult problems for the management of aggregate demand in the importing countries and put pressure on the operation of the international trading and financial system as well.

This study is concerned with the way in which the world economy responded to these pressures in 1974 and with the probable course of adjustment during the period 1975–77. There are four main strands running through the analysis. First, the most serious problems caused by the sharp rise in oil prices were the immediate fall in aggregate demand and the impetus to cost-push inflation generated in the industrial countries. These in turn hurt the economies of the developing countries. Second, the demand-depressing effects of the oil price increase will continue, but will grow steadily smaller in size over the next several years. As this happens, the problems caused initially by higher oil prices for the management of aggregate demand and for the achievement of a satisfactory equilibrium in international payments will gradually give way to the problems associated with the transfer of real resources from oil-importing to oil-exporting countries. The two sets of problems cannot exist to the same degree at the same time; the shift from the one to the other reflects the difference between the immediate and long-term effects of high oil prices. Third, the repercussions of the oil crisis, large as they are, are felt within a world economy that is enormous in size and possesses—at least potentially—a considerable capacity to absorb change. Fourth, the realization of this capacity depends principally on the accuracy with which the oil problem is diagnosed, the appropriateness of the domestic policy responses, and the effectiveness with which policies are co-ordinated by the United States, Japan, and the countries of Western Europe.

The analysis is based on a series of papers prepared by several authors and discussed at a conference held at the Brookings Institution 18–20 November 1974. The authors employed a variety of techniques for measuring the impact of high oil prices on the several countries, regions, and sectors examined, but the analyses are in general consistent with each other. These papers have since been revised in the light of the discussion that took place at the conference and have been brought up to date to the extent that it was possible to do so. In the rest of this chapter the results are summarized from a global point of view, drawing on the rest of the book whenever it is appropriate. In addition, this chapter includes a discussion of long-run welfare costs associated with high oil prices and outlines the policy implications of the study as a whole. In Chapters 2 through 6 it is assumed that the present level of oil prices will remain unchanged in real terms through the period 1974–77, and the effect of these prices on the economies of the United States, Western Europe, Japan, and the developing countries and on the operation of the international financial system is estimated. The final chapter contains an analysis of the factors determining the volume of world oil exports, on the basis of which the authors examine possible marketing strategies that might be adopted by the Organization of Petroleum Exporting Countries (OPEC) and other considerations that will influence the actual course of oil export prices over the medium term.

The Nature of the Adjustment Process

Because they operate in such different economic environments, it is helpful to consider the industrial countries and the nonoil developing countries (that is, the oil-importing developing countries) separately in analyzing the effects of the oil price increase.

In the case of the industrial countries the adjustment process can be thought of as taking place in several distinct phases: *First,* an initial phase, during which the large increase in the price of oil raises the general price level and simultaneously transfers income from consumers to producers of energy, who in turn accumulate a large fraction of their suddenly swollen incomes in unspent financial surpluses. *Second,* a transition phase in which producers of energy gradually increase their spending out of the higher receipts; oil-exporting countries increase their purchases from oil-importing countries; and producers of energy within the consuming countries expand their production facilities in response to the higher prices of their products. *Third,* a final phase in which the transition is completed, and consumers of energy are fully paying for the higher prices through a transfer of real resources—as reflected in higher exports to foreign producers of energy and higher resource costs for domestic production of primary sources of energy.

The economic consequences of higher oil prices for the industrial countries are quite different during the various stages of this process and hence require different economic policy responses. In essence, the initial impact of the oil price increase can be compared to the imposition by the producers of oil of a large excise tax, the proceeds of which were *not* immediately used to buy goods or services. Consumers in the importing nations paid more for energy and therefore had less to spend for other products. While exports to the oil-producing countries expanded, the increase in 1974 absorbed only a small portion of their increase in oil revenues. There was not, moreover, a sharp rise in domestic energy investment—at least not initially. As a consequence, sales, output, and employment were reduced in the consumer-goods industries. Total demand and output fell, not only by the amount of the initial loss in consumer demand, but still further through the typical cyclical process in which the initial reductions in employment and income were the cause of still further declines in demand, in output, and in jobs. Moreover, the loss of output and income in each country tended to be self-reinforcing, since each country provides an export market for the others.

At the same time that the excise tax on oil depressed economic activity it added to the pace of inflation, which before the October embargo had already been accelerating in most parts of the world. The sharp increase in the price of crude oil was passed through to the prices of gasoline, heating oil, electricity, petrochemicals, and other petroleum-using products. Prices of domestically produced fuels increased partly or fully in line with imported oil prices. Wage increases accelerated in response to the higher prices, leading to additional price increases in areas outside the energy industries. Cost-push inflation was thus given a substantial added thrust.

Initially, therefore, the unprecedented jump in oil prices raised major problems for the management of aggregate demand, setting in motion large recessionary forces and giving another turn of the screw to cost-push inflation. In early 1975 these initial impacts still predominated, but the second, transitional, phase had begun to set in. So long as oil prices remain at or above the $10 level, the drain of consumer purchasing power will continue. But its impact on aggregate demand should steadily be moderated by several developments: Demand for oil and other energy products should be increasingly restrained by the higher prices. The response (elasticity) of total energy consumption to higher prices and the substitution of other fuels for oil occur gradually. If world oil prices remain at the $10-plus level of late 1974, total energy consumption should fall increasingly below the path it would have taken in the absence of the price increases. Even more will this be true of imported oil, as increased domestic production of primary sources of energy is substituted for oil purchased abroad. The drain on purchasing power should therefore be reduced. Further, exports to the oil-producing countries will rise, absorbing a steadily larger fraction of those countries' oil receipts. Finally, to a lesser extent—and subject to some major uncertainties discussed below—investment in the expansion of domestic sources of energy can be expected to increase in Western Europe and the United States. (Japan has few domestic energy alternatives to exploit.)

The impact of the 1974 oil price increase on the general rate of inflation in the industrial countries has probably run most of its course already.[1] The prices of goods and services which use petroleum or competing sources of energy have, by now, been marked up to reflect most, if not all, of the increased costs. It is more difficult to generalize about indirect effects on subsequent wage negotiations. The extent to which increases in consumer prices are automatically passed through into higher wages varies substantially among the industrial countries. Because of the oil price increase, consumer prices in 1975 will be higher than they would otherwise have been, and future wage contracts may well reflect some of this increase. On the other hand, because of recession and falling prices for other raw materials and agricultural products, inflation is abating in most industrial nations. It is virtually impossible, at this stage, to disentangle all the forces operating on the setting of wage rates, to determine whether earlier increases in consumer prices attributable to higher oil prices are any longer having significant effects in pushing up wages and other prices. On balance, however, further inflationary effects of the 1974 oil price rise are likely to be quite small.

Assuming that the real price of imported oil remains the same, therefore, the next few years should see a transition phase in which the problems associated with the unemployment- and inflation-creating aspects of

---

1. This is not true, however, for the United States. See below, p. 28.

the oil price rise will gradually give way to problems associated with the transfer of real resources to the oil-exporting countries and to domestic producers of energy. The higher oil prices will lead to a slightly reduced rate of growth in living standards, not because output and employment are still curtailed but because each unit of energy consumed will cost more in terms of economic resources. Adjusting to higher oil prices with minimum economic loss will no longer depend on optimum policies for managing aggregate demand but on the measures chosen to affect the supply and demand for energy products themselves.

In the case of the nonoil developing countries the problems of higher oil prices center chiefly around their impact on foreign-exchange earnings and reserves, rather than on aggregate demand. Economic growth in most of these countries is heavily dependent on the availability of foreign exchange, which in turn is a major factor determining their capacity to import capital goods, to support investment, and to generate economic growth. They have been and will continue to be affected by higher oil prices in four principal ways: *First,* the oil-induced recession in the industrial countries shrinks their export markets and thereby reduces their ability to import. *Second,* out of their reduced foreign-exchange earnings a larger fraction has to be devoted to paying for oil, leaving a smaller amount available for the other imports needed to meet development plans. *Third,* adverse changes in their trade balances can impair the ability of developing countries to borrow in private capital markets. And *fourth,* the transfer of income from OECD countries to members of OPEC can affect the flow of concessional aid to the developing countries.

Working Assumptions

In outlining the adjustments confronting the oil-importing countries, we start with estimates of the magnitude of the initial shocks to aggregate demand and the way in which these might change over the medium term. Six major sources of shock or economic impact can be distinguished:

• The additional cost of imported oil: the size of the impact depends on the relative importance of imported oil in each nation's economy as well as on the size of the price increase.

• The increase in exports from countries of the Organisation for Economic Co-operation and Development (OECD) to the oil-exporting countries because of the improvement in the financial position of the latter. The more the increased costs of imported oil are returned in the form of additional exports to the oil-producing countries, the less will higher oil prices depress aggregate demand in the oil-importing countries.

• The reduction in OECD exports to the oil-importing (or nonoil) developing countries. The latter will have to reduce their imports from the OECD countries because their export earnings are adversely affected by the world recession and because they must use a larger proportion of their available foreign exchange to pay for imported oil.

• Changes in the trade of the OECD countries with each other induced by the internal effects of higher oil prices in reducing gross national product (GNP).

• The impact of higher oil costs and the oil-induced reductions in world GNP on the trade and income of the developing countries, as modified by offsetting or aggravating changes in capital movements.

• Increased prices paid in the importing countries for domestically produced oil and other primary sources of energy.

Working assumptions about the first three of these sources of impact will be outlined in this section, while the fourth, fifth, and sixth will be discussed in the following two sections as part of the analysis of developments in the industrial countries and in the developing countries.

For present purposes, the impact from the first three sources is estimated to be the difference in developments under two projections: first, what might have been expected to occur in the pre-October 1973 environment, under the assumption of a continuation of economic growth at more or less trend rates and with oil priced at $2.75 a barrel, and second, what might be projected in mid 1975 with the world economy in recession and with oil priced at $10 a barrel. Thus for each of the three exogenous variables we are considering here—oil import costs, OECD exports to the oil-exporting countries, and OECD exports to the oil-importing developing countries—present projections are compared with a base projection to arrive at a usable definition of incremental change.

Estimating the additional costs of imported oil is of course the key calculation. For this purpose we project as a base case that given the conditions prevailing before October 1973 the oil imports of non-Communist countries would have increased from 30 million barrels a day (MBD) in 1973 to 42 MBD in 1977, or by approximately 11 percent a year.[2] In contrast, our projection in present circumstances, as developed by Joseph Yager and Eleanor Steinberg in Chapter 7, shows an absolute decline in oil-import requirements in both 1974 and 1975; only by 1977 do imports regain their 1973 level. The two projections are compared in Table 1-1.

These data suggest that as a result of both the worldwide recession and higher energy prices oil import requirements in 1977 could be down by 12 MBD, or almost 30 percent, from the base or trend level. Of this decline, approximately 4.5 MBD can be attributed to the effect of the recession on the demand for energy, 4 MBD to the effect of higher oil prices on the demand for energy, and 3.5 MBD to the effect of higher oil prices in stimulating production of all primary sources of energy in the oil-importing countries.[3]

Even so, these oil-import projections are derived from essentially conservative or pessimistic assumptions about the effect of higher oil prices on the consumption of energy and the production of primary sources of energy in the oil-importing countries. On the demand side, the consumption response to the sharp increase in price is assumed to be comparatively small. (Specifically, the price elasticity for total consumption of energy underlying the present projection is assumed to be the same for all industrial countries and to grow in equal annual increments from -0.03 in 1974 to -0.12 in 1977.)

On the demand side, two aspects of these projections require special note. Contrary to what one might expect, U.S. oil imports show the smallest change from the base projection and hence the smallest response to the dramatically changed situation in energy. This result follows from the pessimistic assumptions deliberately employed by Yager and Stein-

2. This rate of increase is based on projections appearing in *Energy Prospects to 1985: An Assessment of Long Term Energy Developments and Related Policies,* 2 vols. (Paris: Organisation for Economic Co-operation and Development, 1975). The calculations are explained in the footnotes to Table 1-1.

3. The estimated effect of the recession on the demand for oil is based on the difference between presently projected and trend growth rates multiplied by the oil-GNP relationship shown in Chapter 7, page 255. The estimated effect of the higher price of oil on the demand for oil is that shown in Table 7-8. The estimated increase (from trend levels) in domestic production of primary sources of energy accounts for the remaining difference. (Since the recession in OECD countries, especially in the United States and Japan, has been larger than can be explained by the oil crisis, the recession-induced reduction in imports of oil cannot be attributed solely to the direct and indirect influence of the rise in oil prices.)

Table 1-1. *Projected Oil-Import Requirements of Non-Communist Countries, 1974–77*[a]

Billions of barrels

Year and projection	United States	Western Europe	Japan	Other industrial countries	Developing countries	Total
1973	6.4	14.4	5.3	0.5	4.1	30.7
1974						
Pre-October 1973 projection	7.2	15.3	5.7	0.6	4.5	33.3
Present projection	6.1	13.6	5.0	0.5	4.0	29.2
1975						
Pre-October 1973 projection	8.1	16.2	6.2	0.7	4.8	36.0
Present projection	5.8	12.8	4.8	0.6	3.9	27.9
1976						
Pre-October 1973 projection	9.3	16.8	6.8	0.7	5.2	38.8
Present projection	6.8	12.3	4.7	0.6	4.1	28.5
1977						
Pre-October 1973 projection	10.5	17.4	7.4	0.8	5.7	41.8
Present projection	8.0	11.7	4.9	0.7	4.4	29.7

a. Import requirements exclude adjustments for changes in stocks.

Pre-October 1973 projections assume trend growth rates and a price of $2.75 a barrel for oil. They are estimated as follows:

United States: Import requirements for 1973 are increased by 13.7 percent a year. This rate of increase is based on the difference between 1973 imports and OECD projections of imports in 1980 at a price of $3 a barrel (OECD, *Energy Prospects to 1985*, Vol. 1, p. 12), adjusted to exclude production from Alaska and to take into account more recent information on domestic energy production (see Chapter 7). The figure thus obtained for 1977 is reduced by 0.1 MBD to allow for the initial operation of the Alaskan pipeline in the fourth quarter of that year.

Western Europe: Import requirements for 1973 are increased by 6.3 percent a year. This rate of increase is based on the difference between 1973 imports and OECD projections of imports in 1980 (*Energy Prospects to 1985*), adjusted to exclude production from the North Sea. Resulting figures thus obtained for 1975, 1976, and 1977 are then reduced respectively by 0.1 MBD, 0.5 MBD, and 1.0 MBD to allow for estimated North Sea production in those years.

Japan: Import requirements for 1973 are increased by 9.2 percent a year. This rate of increase is based on the difference between 1973 imports and OECD projections of imports in 1980 (*Energy Prospects to 1985*).

Other Industrial Countries (Australia, New Zealand, and South Africa): Actual imports in 1972 are increased by 6.9 percent a year (based on long-term projections by Edward R. Fried presented in Joseph A. Yager, Eleanor B. Steinberg, and others, *Energy and U.S. Foreign Policy* (Ballinger Publishing Company, 1974). Import requirements include oil supplied for bunkers, on the assumption that the amount of such oil is roughly equal to the amount supplied elsewhere for incoming transportation.

Developing Countries: Actual imports in 1972 are increased by 8.3 percent a year (based on long-term projections by Edward R. Fried, in Yager, Steinberg, and others, *Energy and U.S. Foreign Policy.*) Import requirements include oil supplied for bunkers, on the assumption that the amount of such oil is roughly equal to the amount supplied elsewhere for incoming transportation.

Present projections are those shown in Chapter 7. They assume that the price of oil will be $10 a barrel (in 1974 dollars) throughout the period.

berg in calculating U.S. production of primary sources of energy—specifically their projection of a steady decline in the production of oil and natural gas and only moderate increases in coal and nuclear power. As a consequence, U.S. production of all primary sources of energy in 1977 is projected to be down slightly from the 1973 level despite an almost threefold increase in domestic prices. Another reason for the relatively high projection of U.S. oil imports in 1977 is that only a minimal allowance was made for Alaskan oil, since the pipeline is not expected to come into large-scale operation until 1978.[4]

On the other hand, Western European imports show a strikingly large response to higher oil prices, accounting for more than half the estimated reduction from trend in world oil imports. This is principally the consequence of substantial increases in production of natural gas and nuclear power and a steady increase during the period in production of oil from the North Sea—all of which is consistent with available information about present programs and plans. The contrast between U.S. and Western European trends reflects in part the fact that even before the price increase U.S. oil imports were expected to rise much more rapidly than those of Western Europe. Nonetheless, to the extent that a bias exists in these projections the consequences are likely to be partially offsetting—that is, the projection of U.S. oil imports in 1977 may prove to be too high and that for Western Europe too low.[5]

The extent to which higher oil prices add to the cost of oil imports can be calculated directly from these volume projections, as shown in Table 1-2. It is evident that the effect of the sharp increase in price, from $2.75 a barrel in 1973 to $10 a barrel in 1974 and thereafter, overwhelms the effect of the reduction in volume, substantial though the latter is when

4. In addition, Yager and Steinberg's projections are based on a continuation of 1974 energy prices in the United States. The average price of fossil fuels in the United States is controlled at levels below world prices. The impasse between the President and the Congress on the future of controls was still in effect at the time of writing (mid September 1975). Whatever the final outcome may be, it is likely to include some net increase over 1974 prices. On both the demand and supply sides, this increase should operate to yield some reduction in imports, even within the period under consideration.

5. Projections for Western Europe, however, are also conservative in that they are based on average 1974 prices for primary sources of energy. Hence they make no allowance for the effect on energy consumption of additional increases that have occurred in excise taxes on gasoline and in real prices for coal and natural gas.

Table 1-2. *Incremental Costs of Oil Imports, 1974–77*
Billions of 1974 dollars

Year	United States	Western Europe	Japan	Other industrial countries	Developing countries	Total
1974	15.0	34.3	12.5	1.2	10.1	73.1
1975	13.0	30.5	11.3	1.5	9.4	65.7
1976	15.5	28.0	10.3	1.5	9.7	65.0
1977	18.7	25.2	10.4	1.8	10.3	66.4

Source: Table 1-1. Figures may not add to totals because of rounding.
Incremental costs represent the difference between pre-1973 projections of import requirements at $2.75 a barrel and present projections of import requirements at $10 a barrel. Production costs and company profits are assumed to be constant. In these figures, changes in stocks are disregarded; specifically, no allowance is made for the substantial increase in stocks that took place in 1974 or for the reduction of stocks that is taking place in 1975.

compared with trend expectations. As a result, the incremental cost of oil imports is estimated to have been almost $75 billion in 1974 and is projected at approximately $65 billion a year for the following three years. For the United States the incremental cost declines in 1975 and then rises rapidly in 1976 and 1977, on the assumptions of a sharp economic recovery and, as emphasized earlier, slow responses of domestic production of primary sources of energy to higher prices. In Western Europe, on the other hand, the substantial increases in domestic production of primary sources of energy cause a steady decline in the costs of incremental oil imports.

Against these additional foreign-exchange costs the oil-importing countries can expect to increase their sales of goods and services to the oil-exporting countries. The rapidity with which this happens will depend on the capacity of the oil-exporting countries to use additional imports of goods and services for domestic consumption and investment (including military programs), the distribution of such imports between those that are delivered more or less immediately and those requiring long lead times, and the priorities attached to such expenditures. Since the oil-exporting countries differ in these respects, the expenditures of the group as a whole will also depend on the way in which the export market for oil—and hence the additional oil revenues—are distributed among them. For example, the greater the share enjoyed by countries with a large capacity to absorb imports, the more rapidly incremental oil revenues will be spent, and vice versa.

We have projected in Table 1-3 a pattern of such expenditures that takes these factors into account. It is based on the following suppositions:[6]

• Those oil-exporting countries with low per capita incomes and those accounting for a relatively small share of the market (most of which are not members of the Organization of Petroleum Exporting Countries [OPEC]) will increase their exports pretty much in proportion with the growth in their productive capacity. The countries in this group—Algeria, Indonesia, Nigeria, the USSR, China, and fifteen other countries with relatively small exports—are projected to account for an average of one-fourth of total world exports during this period. These countries have large absorptive capacities and normally are substantial borrowers in world markets. It is assumed that the group as a whole will commit its entire incremental oil receipts each year for imports of goods and services and, conservatively, that these commitments will be spent in equal amounts over a three-year period.

• A second group of oil-exporting countries, consisting of Iran, Iraq, and Venezuela, will accept substantial responsibilities for cutting back production when this is necessary to keep exports in balance with world demand at a price of $10 a barrel. Consequently, the oil exports of these countries, while accounting for one-third of the world oil market, will be approximately 40 percent below production capacity. These countries, Iran in particular, have undertaken rapidly expanding economic and military programs. It is assumed further that the group as a whole will commit its entire incremental oil revenues each year to imports of goods and services. Since these commitments will be very large and since a substantial proportion will be used to finance industrial projects and other programs with long lead times, it is assumed that they will be spent over a five-year period.

• A third group of oil-exporting countries, consisting of Kuwait, Libya, Qatar, Saudi Arabia, and the United Arab Emirates, will also restrain exports as part of their responsibilities as residual suppliers. Their oil exports, while accounting for 40 percent of the world market, will be little more than half their productive capacity. Countries in this group would not be able to increase their commitments to expenditures by an amount equal to the increase in their oil revenues, both because their eco-

6. For an elaboration of these suppositions based on the situation in each of the major oil-exporting countries, see Yager, Steinberg, and others, *Energy and U.S. Foreign Policy*, pp. 282–89.

Table 1-3. *Expenditures of Oil-Exporting Countries*
on Imports of Goods and Services, 1974–77
Billions of 1974 dollars

Item or country	1974	1975	1976	1977
	Revenues and expenditures			
Total oil revenues[a]	*107.4*	*102.9*	*106.5*	*111.9*
Normal oil revenues[b]	33.3	36.1	38.9	41.9
Additional oil revenues[e]	74.1	66.8	67.6	70.0
Total imports	*42.6*	*57.2*	*74.0*	*86.5*
Normal expenditures[d]	30.3	33.1	35.9	38.9
Additional expenditures[e]	12.3	24.1	38.1	47.6
	Apportionment of additional expenditures among nonoil-exporting countries[f]			
United States	2.6	5.1	8.1	10.1
Western Europe	5.3	10.4	16.5	20.6
Japan	2.3	4.5	7.0	8.7
All other	2.1	4.1	6.5	8.1

a. Present projection of oil imports (Table 1-1) at $10 a barrel plus interest income on the accumulating financial balances amounting to $0.8 billion in 1974, $1.1 billion in 1975, $2.5 billion in 1976, and $3.5 billion in 1977. (The interest rate is assumed to be 3 percent a year in real terms.)

b. Pre-1973 projection of oil imports (Table 1-1) at $2.75 a barrel.

c. Difference between a and b.

d. Normal oil revenues less allowance for current-account surplus of $3 billion a year (1973 level). Does not include imports financed by nonoil exports; these are assumed to be equal and therefore to have no effect on additional expenditures.

e. Based on allocation of projected world oil export market as follows:
Countries spending incremental revenues over a three-year period (Algeria, Indonesia, Nigeria, smaller exporting countries, USSR, and China). Exports of these countries are projected to be 6.4 MBD in 1974, 6.7 MBD in 1975, 7.8 MBD in 1976, and 8.6 MBD in 1977.
Countries spending incremental revenues over a five-year period (Iran, Iraq, and Venezuela). Exports of these countries are projected to be 9.6 MBD in 1974, 9.6 MBD in 1975, 9.7 MBD in 1976, and 9.9 MBD in 1977.
Countries increasing expenditures on imported goods and services by 20 percent a year (Kuwait, Libya, Qatar, Saudi Arabia, and the United Arab Emirates). Oil exports of these countries are projected to be 13.2 MBD in 1974, 11.6 MBD in 1975, 11.0 MBD in 1976, and 11.2 MBD in 1977. Normal expenditures of these countries are assumed to be equal to their normal oil receipts less a current-account surplus of $3 billion a year. Their additional expenditures are the difference between total imports and services, projected to grow by 20 percent a year from the 1973 level, and normal imports.

f. Market shares are based on the sources of supply for the *increase* in OPEC imports during 1974, (*UN Monthly Bulletin of Statistics*, June 1975, Special Table B).

Figures may not add to totals because of rounding.

nomic base is too narrow and because their populations are too small. Nonetheless, as events in 1974 have shown, their expenditures to increase consumption immediately and to support ambitious economic development and military programs are growing very rapidly; for the group as a whole it is assumed that expenditures on imports or goods and services will increase by 20 percent a year over the period 1975–77.

Under these assumptions, the oil-exporting countries would spend a rapidly increasing share of their *incremental* revenues on imports of goods and services, with the proportion rising from one-sixth in 1974 to more than two-thirds in 1977. Most of these expenditures—about 85 percent—would be made in the United States, Western Europe, and Japan, because the OECD countries normally are the predominant sources of supply for the oil-exporting countries and because these incremental expenditures would be heavily concentrated on industrial and military equipment and technical services. These results, it should be stressed, would not be substantially different with prices moderately higher or moderately lower than $10 a barrel. At a higher price, the volume of oil exports over the medium term might be reduced by somewhat less than the increase in prices, thus leading to even larger incremental revenues. On the other hand, a disproportionate part of the cut in exports would have to be borne by the countries that have low absorptive capacities, so total expenditures of the oil-exporting countries might increase even more rapidly than has been projected. At moderately lower oil prices the opposite situation could exist. Thus, in most conceivable circumstances, the initial external shock effects imposed on the OECD countries by higher oil prices will be moderated by their accelerating exports to the oil-producing countries. As this happens, the consequences of higher oil prices for aggregate demand and the balance of payments will be replaced by those having to do with the transfer of real resources.

The third exogenous variable to be considered in this section is the negative effect of higher oil prices on OECD exports to the oil-importing developing countries. Simply stated, the latter will have to reduce their imports from the OECD countries because the volume of their exports, their terms of trade, and hence their buying power, will suffer from the oil-induced recession in the industrial countries and because they will have to spend more of their available foreign exchange on imports of oil and less on imports of industrial and other products. This reduction in their purchases will be mitigated to the extent that these countries can increase their capital imports from both the OECD and the oil-exporting countries, something which did indeed happen in 1974.

Our projection of these OECD export losses is drawn from the econometric model of the International Bank for Reconstruction and Development (IBRD, the World Bank) that links the trade of the developing countries with economic growth in the industrial regions. It represents the difference between a projection of OECD exports to the oil-importing developing countries in pre-October 1973 circumstances (that is, assuming normal OECD growth trends and oil at $2.75 a barrel) and a present

Table 1-4. *Projected Reductions from Pre-October 1973 Trend Levels in OECD[a] Exports to Oil-Importing Developing Countries[b]*
Billions of U.S. dollars

Country or region[c]	1974	1975	1976	1977
United States	0.4	2.1	2.9	3.2
Western Europe	0.5	2.9	4.0	4.4
Japan	0.3	1.7	2.3	2.6
Total	1.2	6.7	9.2	10.2

a. Excluding Australia, Canada, and New Zealand.
b. The basis for reductions in the imports of oil-importing developing countries from pre-October 1973 trend levels, and hence of the reductions in OECD exports, is explained in the text.
c. The apportionment of the reduction in exports from pre-October 1973 trend levels among the OECD countries is based on market shares for trade in 1974 (*UN Monthly Bulletin of Statistics*, June 1975).

projection based on current economic prospects and oil at $10 a barrel. The results, shown in Table 1-4, suggest that the reduction in OECD exports on this account might amount to $1 billion in 1974 and increase to about $10 billion in 1977.[7]

Thus the assumptions we have made about the probable course of the world oil market and the world economy and about the expenditure responses of the oil-exporting countries to the increase in their revenues would indicate that the external shock effects of higher oil prices on the OECD countries as a group will decline substantially over the next few years. The calculations based on these assumptions are summarized in Table 1-5. They show that by 1977 the aggregate effect of the three factors we have considered will be about half their initial size. For the United States, however, little change will have taken place because of deliberately pessimistic assumptions about domestic energy production, which lead to assumptions of high oil import requirements. In Western Europe, on the other hand, the external impact of higher oil prices on aggregate demand

7. These figures reflect the effects of all projected changes in the world economy (including projections of changes in commodity prices and of the nonoil terms of trade) and hence are larger than the amount that might be attributed to higher oil prices alone. Losses were small in 1974 because of unusual gains realized by these countries in 1973 from high commodity prices, gains that were used in part to add to foreign-exchange reserves rather than for imports; because of the offsetting effects of an increase in capital flows; and because of lags in some of the economic responses to recession. In 1975 and for several years thereafter, capital flows to the developing countries are projected to decline in real terms, and the recession in OECD countries is projected to shift the nonoil terms of trade against the oil-importing developing countries, thus further restricting their available foreign exchange.

Table 1-5. *Summary of Oil Impact Assumptions for the United States, Western Europe, and Japan, 1974–77*

Billions of 1974 dollars; signs indicate direction of effect on aggregate demand

Country or region and assumption	1974	1975	1976	1977
United States				
Increased cost of oil imports	−15.0	−13.0	−15.5	−18.7
Reduced exports to oil-importing developing countries	−0.4	−2.1	−2.9	−3.2
Increased exports to oil-exporting countries	+2.6	+5.1	+8.1	+10.1
Net impact	−12.8	−10.0	−10.3	−11.8
Western Europe				
Increased cost of oil imports	−34.3	−30.5	−28.0	−25.2
Reduced exports to oil-importing developing countries	−0.5	−2.9	−4.0	−4.4
Increased exports to oil-exporting countries	+5.3	+10.4	+16.5	+20.6
Net impact	−29.5	−23.0	−15.5	−9.0
Japan				
Increased cost of oil imports	−12.5	−11.3	−10.3	−10.4
Reduced exports to oil-importing developing countries	−0.3	−1.7	−2.3	−2.6
Increased exports to oil-exporting countries	+2.3	+4.5	+7.0	+8.7
Net impact	−10.5	−8.5	−5.6	−4.3
Total				
Increased cost of oil imports	−61.8	−54.8	−53.8	−54.3
Reduced exports to oil-importing developing countries	−1.2	−6.7	−9.2	−10.2
Increased exports to oil-exporting countries	+10.2	+20.0	+31.6	+39.4
Net impact	−52.8	−41.5	−31.4	−25.1

Sources: Tables 1-2, 1-3, and 1-4.

will drop steadily and sharply throughout the period. For Japan, the pace of adjustment is about midway between that of the United States and that of Western Europe.

The Industrial Countries

Initially, as was noted above, the most widely feared aspect of the October 1973 oil crisis was the sharp reduction of oil exports and the selective embargo by the Arab members of OPEC against the United States and the Netherlands. Other countries such as Japan, not themselves embargoed but heavily dependent on imported oil, began to worry about the security of their own oil supplies. While physical supply shortages caused major inconveniences briefly in some countries, led to various emergency measures to ration scarce supplies, and had a particularly depressing effect on sales of automobiles and houses in the United States, the continuing—and by far the most costly—consequences of the oil crisis came as a result of the quadrupling of world oil prices.

At its peak, in the winter of 1973–74, the reduction in oil exports by the Arab members of OPEC amounted to about 10 percent of world oil consumption. Although the Arab exporters subsequently sought to make the United States and the Netherlands the principal objects of the embargo, data for the first quarter of 1974 indicate that the international oil companies spread the shortage among the United States, Japan, and Western Europe roughly in proportion to their consumption of petroleum. The reduction of exports to the United States may have been somewhat greater than the average, but not much greater. In any event, by the spring of 1974 purely physical shortages had disappeared.

The direct economic effects of this reduction in oil supplies were relatively modest. In most of the industrial countries policies were adopted which prevented a shortage of petroleum from curtailing industrial production and employment to any significant degree. Cutbacks in oil consumption were concentrated heavily on gasoline and, to a lesser extent, heating oil. The United States, with its proportionately larger consumption of gasoline, was probably in a better position to contain economic disruption by such allocation measures than were Western Europe and Japan.

Embargo-induced fears and uncertainties about the future availability of petroleum products, however, had much more significant effects on government economic policies and consumer behavior in the industrial countries and hence on their economies. In Chapter 2, George Perry estimates that automobile sales in the United States were reduced by $5.8 billion during the year beginning in October 1973, by a combination of higher gasoline prices and uncertainties about shortages. Home building in the United States, already declining under the influence of higher interest rates, was probably reduced further by the same combination of cost and psychological factors. In Japan, initial estimates made in late 1973 predicted a 20 percent reduction in the availability of oil. While such a

Table 1-6. *Relative Importance of Energy, Oil, and Imported Oil in the Three Regional Economies, 1973*

Country or region	Barrels of oil or oil equivalent per thousand dollars of GNP			Index (U.S. consumption per thousand dollars of GNP = 100)		
	Total energy	Oil	Imported oil	Total energy	Oil	Imported oil
United States	10.18	4.79	1.72	100	47	17
Western Europe	6.51	4.10	4.00	64	40	39
Japan	6.19	4.75	4.75	61	47	47

Sources: Energy data from Tables 7-5, 7-6, and 7-7 of this volume. 1973 GNP from official national income statistics.

shortage never occurred, fear of its consequences in the form of reduced industrial production contributed to the imposition of excessively tight monetary and fiscal policies.[8]

As imported oil became freely available, the industrial countries had to face the continuing problems of adjusting to the effect of higher oil prices on aggregate demand and inflation. In the next two sections, relying principally, but not solely, on the regional analyses in Chapters 2, 3, and 4, we summarize the effect in 1974 and then discuss the outlook for the next several years of transition.

The Immediate Effect: 1974

Of the three major regions, the United States is the largest consumer of energy in relation to the size of its GNP. Its total consumption of oil, however, is only slightly larger than that of either Japan or Western Europe, and in 1973 some two-thirds of the oil consumed in the United States was produced domestically (see Table 1-6). Looking solely at the importance of imported oil in the economies of the three regions one would have expected to find that the depressing effects of the quadrupled world price of oil were largest in Japan, slightly less in Western Europe, and much less in the United States.

In fact, however, the magnitude of the initial depressing influence of the OPEC price increase on aggregate demand in each major industrial country depended on a large number of factors. Quantitative estimates for

8. In Chapter 4, Tsunehiko Watanabe points out that so sharp a reduction in oil imports, had it occurred, would indeed have resulted in a drastic curtailment of business investment and very large declines in GNP.

Table 1-7. *Selected Factors Affecting the Magnitude of the Initial Economic Impact of Higher Oil Prices in 1974*

Percentage of GDP; signs indicate direction of effect on aggregate demand

Country or region	Increase in oil import bill[a]	Increase in exports to oil-exporting countries[a]	Increase in payments to domestic oil producers[b]	Loss of exports to oil-importing developing countries[a]	Total
United States	−1.03	+0.18	−0.86	−0.03	−1.74
Western Europe	−2.28	+0.35	...	−0.03	−1.96
Japan	−2.67	+0.49	...	−0.06	−2.24

a. Derived from Table 1-5.
b. Based on George Perry's estimates in Table 2-4.

three of these factors are shown in our impact assumptions. Two of them —the increased payments for imported oil and the loss in exports to the oil-importing developing countries—depressed demand; the third—increased exports to the oil exporters—provided a partially offsetting stimulant to demand. For the United States, a fourth factor was significant—the transfer of purchasing power to domestic producers of oil because of rising prices for domestic oil. Although domestic oil prices rose by less than the increase in the price of imported oil (because of the imposition of price controls), the rise was still substantial. As George Perry points out in Chapter 2, total energy investment in the United States did not rise in 1974, so virtually none of this loss of consumer purchasing power was offset by increased demand in the investment goods industries. Thus, virtually all the transfer represented a net loss of demand.[9] When this fact is taken into account, the impact of these four factors on aggregate demand, calculated as a percentage of GNP, appears to have been about the same in the United States and Western Europe and significantly higher in Japan, as shown in Table 1-7.

Another factor applicable to the United States and Western Europe was the transfer of purchasing power to domestic producers of competing fuels (coal and natural gas) whose prices also rose. This led to an additional loss of consumer purchasing power not offset by increased demand

9. Domestic corporations producing energy may have channeled a small part of the rise in profits to extra dividends, which in turn may have increased consumption by recipients of dividends, thereby providing a small offset to the decline in overall consumer demand.

elsewhere. (None of the authors take this factor into account; its overall impact in *1974* was significant but much less than that of oil.) [10]

Finally, these initial demand-depressing effects were aggravated by the internal and external reactions to them. An initial reduction in purchases by consumers, who are diverting more of their incomes to paying for energy, leads to layoffs in the consumer-goods industries. In turn, the reduced incomes of those laid off or of those working shorter hours cause further declines in production and sales of consumer goods. Manufacturers curtail their investments in inventory and ultimately their investments in plant and equipment. As the income of each nation is reduced that nation tends to import less, thereby reducing the export sales and the GNP of its principal suppliers. Further indirect effects occur through the financial markets. For example, the cost-push inflation induced by higher oil prices tends to raise the demand for cash balances, but if monetary policy is not relaxed to accommodate this need, as it was not in the major OECD countries, interest rates will rise, depressing demand still further.

All the authors of the regional chapters employ large econometric models of the economy either directly or indirectly to estimate the year-by-year effect on GNP of the initial oil-induced effects outlined above. The relationships between the initial shock and the ultimate effect on GNP implicit in the various econometric models are usually called "impact multipliers"—the ultimate percentage change in GNP per 1 percent initial shock. Since the effects are transmitted gradually throughout the economy the impact multipliers become larger in the second and third years after the initial shock. And since the economic structures of the various economies are not identical, the impact multipliers—and hence the declines in GNP—are not the same. During the first year after the initial shock, however, the differences in response are not very large. [11]

While the authors of the three regional papers used different econometric models and somewhat different approaches to isolate the impact of higher oil prices on GNP in 1974, their results for each region bear roughly

10. In the United States the price of most natural gas is regulated, and most coal is sold under long-term contracts.

11. In Table 3-2, Giorgio Basevi reproduces impact multipliers developed from Project LINK, a set of large econometric models covering in an interrelated way the economies of the three major regions. As his data indicate, the multipliers for the United States, Japan, and Western Europe are not very different from one another in the first year. The differences, as estimated by the particular set of models that are incorporated in LINK, widen in the second and third years.

the same relationship to each other as the magnitudes of the initial shocks shown in the last column of Table 1-7, above. The initial shocks to the economies of Western Europe and the United States were slightly less than 2 percent of GNP. Our authors estimate that this shock, after first-year transmission effects, led to a decline of approximately 2.5 percent in GNP. The initial shock to the Japanese economy was larger—about 2.25 percent of GNP—and on the basis of Watanabe's analysis this resulted in a reduction of 4.2 percent in aggregate demand and GNP. In all three regional papers the estimates of the change in GNP induced by higher oil prices are made under the assumption that monetary and fiscal policies remain constant in the face of the oil price rise, not because this is what actually happened but because each author was seeking to isolate the effects of higher oil prices from all the other factors influencing the course of economic events.

In Table 1-8, the oil-induced changes in GNP are compared to the economic outlook generally expected before the October 1973 embargo and to the changes which actually occurred in 1974. In both the United States and Japan, for the year 1974 as a whole, higher oil prices accounted for more than one-half the shortfall in GNP below the previously expected path. Both of these countries experienced actual declines in GNP. In Western Europe, on the other hand, GNP continued to grow. And while there was a shortfall in the rate of growth below earlier expectations, it was slightly less, not greater, than the oil-induced impact.

In the light of these estimates of the extent to which the recession was

Table 1-8. *Changes in Real GNP, 1973–74*
Percent

Country or region	Pre-embargo outlook	Actual	Change from pre-embargo outlook		
			Total	Attributable to oil	Attributable to other factors
United States	+2.6	−2.1	−4.7	−2.5	−2.2
Western Europe	+4.8	+2.3	−2.5	−2.7	+0.2
Japan	+5.2	−1.8	−7.0	−4.2	−2.8

Sources:
Pre-embargo forecasts: For the United States—American Statistical Association and National Bureau of Economic Research, *Median Forecasts of Business Outlook Survey,* August 1973; for Western Europe, OECD, *Economic Outlook,* 14 (December 1973); for Japan, derived from quarterly data and from interpolations of fiscal year data presented in Chapter 4.
Actual GNP from official GNP statistics.
Change attributable to oil from the estimates in Chapters 2, 3, and 4 of this volume.

due to higher oil prices and the extent to which it was attributable to other factors, it is useful to examine the way in which monetary and fiscal policy were managed in each of the three regions.

THE UNITED STATES. In the United States fiscal policy grew progressively tighter throughout 1973 and 1974. Revenues respond sharply to rising money incomes under the U.S. tax structure, as taxpayers with growing money incomes move into higher tax brackets. Rapid inflation consequently increases the average effective tax rate, even if real incomes are not rising. Federal budgetary expenditures did not keep pace with the rising effective tax rate. In fact, by the end of fiscal 1974 (June 30), federal expenditures turned out to be $7 billion lower than had been estimated six months earlier. Taxes, moreover, were not reduced; indeed, as late as October 1974 the administration proposed a tax increase (but shortly withdrew its request). The full-employment balance in the federal budget grew from almost zero in the second quarter of 1973 to $30 billion in the third quarter of 1974.[12] Monetary policy also became quite restrictive, since the Federal Reserve chose not to accommodate the externally generated oil price increases with a more liberal expansion of the money supply. Indeed, the opposite occurred. Between November 1973 and August 1974 the money supply rose by 5.5 percent, while prices rose by 11 percent. The real stock of money fell sharply. Short-term private interest rates rose to a peak of almost 12 percent in August 1974. And during the remaining months of the year, the collapse of private demands for credit resulted in only a modest drop in short-term interest rates as the monetary authorities slowed the growth of the money supply further.

Already confronted with accelerating inflation and then faced with the external shock of quadrupled world oil prices, which simultaneously raised prices and reduced aggregate demand, the makers of economic policy chose to pursue restrictive measures throughout 1974, thereby strengthening the forces that were depressing demand, in order to choke off the wage-response and subsequent second-round price effects. After rising gradually in the first nine months of 1974, unemployment rose precipitously toward the end of the year to a level far exceeding that of any other

---

12. The full-employment balance is determined by calculating revenues and expenditures under existing laws as they would turn out under conditions of full employment. In this way changes in fiscal policy can be isolated from those changes in revenues or expenditures due solely to cyclical fluctuations in national income.

period since the Second World War. By the first quarter of 1975 the rate of price inflation in the United States did begin to recede.

What would have been the results had the economic policy been less restrictive—had demand management been directed more toward maintaining employment and less toward containing inflation? More particularly, how much larger would wage gains, commodity price rises, and subsequent second-round inflation have been?

There is reason to believe that wages in the United States respond less sharply to externally generated price increases than do those in most other countries. But the relationship is not known with any precision. Nor is there any settled view of the specific relationship of wages to changes in the level of unemployment. The two econometric models used by George Perry in Chapter 2, for example, imply quite different wage behavior with respect to simultaneous increases in the cost of living and the rate of unemployment. As a consequence, we cannot with any confidence answer the question posed above—how much second-round inflation was prevented in the United States by the restrictive economic policies and the large increases in unemployment which followed the October 1973 oil crisis.[13] What is clear, however, is that the cost of the oil price increases and the restrictive economic policies together was very high in lost output and employment. By the first quarter of 1975, real GNP in the United States was some 11 percent below the level consistent with a 5 percent unemployment rate, principally on account of oil and economic policy.

JAPAN. In Japan the rise in oil prices impinged on an economy already experiencing double-digit inflation. Unlike past inflationary episodes, moreover, the 1973 inflation extended to prices of manufactured goods, which form the predominant part of Japanese exports. Japanese policy makers feared a substantial acceleration of an already large inflation, not only from the rising price of oil itself but from an expected physical shortage of oil and the consequent emergence of widespread supply bottle-

13. In a paper prepared for another conference (George L. Perry, "Policy Alternatives for 1974," *Brookings Papers on Economic Activity, 1:1975*, pp. 222–35), Perry estimated that a more accommodating monetary policy—one which kept short-term interest rates stable after early 1974—and a $20 billion tax cut would have reduced the unemployment rate by 1 percent from its actual level in late 1974 at the cost of an additional rise of 0.5 percent in prices. These results—especially the consequences for prices—would vary depending heavily upon the structure of the particular econometric model used by Perry.

necks. As a result, the oil "crisis" became the occasion for tightening monetary policy further and adopting restrictive fiscal measures that many observers felt should have been undertaken earlier. Fiscal and monetary authorities were further strengthened in their resolve to tighten economic policy by the 33 percent increase in wage rates negotiated for most Japanese workers in the spring of 1974.

Tsunehiko Watanabe, in Chapter 4, highlights the severity of the Japanese demand-management policies. According to his estimates, wholesale prices by the third quarter of 1974 were some 15 percent higher than they would have been without the oil crisis and the events that followed. Far from accommodating at least part of the externally generated price increases, the rate of growth in the money supply was cut, according to Watanabe's estimates, to levels some 4 to 5 percent a year below the path previously planned. In absolute terms, wholesale prices in late 1974 were 21 percent, and consumer prices 25 percent, above those of a year earlier, while the money supply grew by only 10 percent; the result was a sharp fall in the real stock of money.

By the end of 1974 wholesale prices appeared to have virtually stabilized in Japan, and in the first quarter of 1975 the inflation in consumer prices had dropped to an annual rate of 7 percent. The foreign balance on current account swung from a large deficit in the first half of 1974 ($11 billion annual rate) to a substantial surplus in the fourth quarter of the same year, despite a continued heavy net drain from higher oil import costs. The wage negotiations in the spring of 1975 produced a 13 percent increase, substantially less than had been expected. On the other hand, national output continued to fall. In the fourth quarter of 1974, real GNP was 6 percent, and industrial production 12 percent, below the level of a year earlier, and declines in output continued into early 1975.

By every indicator the inflationary surge of 1973–74—more severe in Japan than in most other countries—was rapidly tapering off in early 1975, and the impact of oil payments on the balance of payments had been swiftly offset. A severely restrictive economic policy, which strengthened the demand-depressing effects of the oil price increase rather than moderating them, was chiefly responsible. The cost in terms of lost output was large. By mid 1974, production in the Japanese economy was about 12 to 14 percent below even a modest projection of its potential growth. Precisely because some restrictive policy measures would have been in order in any event, it is difficult to judge how much more output had to be sacrificed to the anti-inflationary objective on account of the increase in oil prices. It is clear from Professor Watanabe's estimates, however, that the additional costs were substantial.

WESTERN EUROPE. For 1974 as a whole, the Western European economy behaved more favorably than that of either the United States or Japan. While the depressing influence of higher oil prices on aggregate demand was about the same as in the United States, real GNP in Western Europe continued to grow in 1974. The growth rate was lower than that expected before the oil crisis, but the reduction from trend was slightly less than could be accounted for by the effect of higher oil prices (see Table 1-8).

The relatively favorable experience for Western Europe, however, results partly from the use of the entire year 1974 as the basis for the comparison. In contrast to the situation in the United States and Japan, where the peak of industrial production was reached in the final quarter of 1973, industrial production in Western Europe as a whole continued to rise in the first half of 1974 and only thereafter began to decline.

The experiences of the individual Western European countries were quite varied. Germany suffered far less inflation than most other nations and ran a strong balance-of-payments surplus. In early 1974, monetary and fiscal policy were relaxed, but output fell gradually through most of the year, turning down sharply near the end. In France, output continued to rise through the first half of the year. Monetary and fiscal policy, which had been restrictive, became even more so after midyear, producing a downturn in production by the year's end. Italy, which suffered from very large balance-of-payments deficits and strongly accelerating inflation, adopted a sharply restrictive economic policy after midyear. Industrial production, which had risen rapidly in the first six months, then turned downward. In the United Kingdom, production was curtailed in the late winter of 1973–74 by a strike of coal miners that forced a three-day work week on the economy. Although production recovered quickly from the slowdown of early 1974, it began falling again in the last months of the year. The smaller countries of Western Europe did better on the whole than the larger ones: their production in 1974 rose by more than 3 percent.

Monetary policy in Western Europe was generally restrictive in 1974, but on balance it was somewhat less so than in the United States and Japan, at least in the first half of the year. Fiscal policy, on the other hand, varied substantially from country to country. In Italy and France, a sharp tightening of fiscal policy occurred after midyear. In the United Kingdom initial intentions to tighten policy were reversed as the year wore on. And in Germany some slight fiscal expansion took place, although major measures to relax fiscal policy came only in 1975. In no country, however, did the government budget swing so dramatically toward restraint as in the United States. In part this may reflect different policy judgments, but more likely it stems from the coincidence that European tax systems are not so heavily weighted toward the progressive income tax as is that of the United States. As a consequence government revenues are far less sensitive to inflation—that is, tax receipts do not rise faster than prices, as they do in the United States. Indeed, given the general desire in early 1974 to restrain demand, it is possible that West European governments might have acted much as the United States government did in allowing the full-employment surplus to accumulate had they possessed a revenue system with characteristics similar to that of the United States.

Despite varying circumstances in the several countries of Western Europe, therefore, a general pattern did appear. During the first half of the year aggregate demand continued, with some moderation, the upward path of the 1972–73 expansion. The demand-reducing effects of the oil price increase were little in evidence, while the price-raising effects, added to the existing inflation, were dominant. In effect, the demand-depressing consequences of higher oil prices, to the extent that they were recognized, appeared as a needed corrective to reduce inflationary conditions of demand. But the combination of increasingly restrictive monetary policies, the demand-reducing effect of the oil price increase and, in some cases, a tightening of fiscal policy soon produced a reversal of gains in output and, by the end of the year, a recession. At the risk of oversimplifying a complex Western European pattern, it may be suggested that the combined effect of the oil crisis and policy measures on aggregate demand in Western Europe was broadly similar to that in the United States and Japan, but on a somewhat smaller scale and with a delay of six months.

As in the United States and Japan, the fall of aggregate demand in Western Europe slowed the growth in wholesale prices by the end of the year. In early 1975 some moderation in the inflation of consumer prices could also be discerned, but to a much smaller degree. The United Kingdom was the exception, as both wholesale and consumer price inflation accelerated sharply in the latter part of the year. What might have been a relatively smooth transition in most of Western Europe from the boom and inflation of 1973 to a more stable pattern of growth was made significantly more difficult and more costly in terms of lost output because of the huge increase in the price of oil.

In sum, the price-raising effect of the oil crisis, following hard on an already accelerating inflation, attracted far more attention from the makers of economic policy in the industrial nations than did its depressing effects on aggregate demand. Monetary policies, far from being relaxed to offset the demand-reducing effects of the oil "excise tax," were tightened in an effort to moderate its inflationary impact. The tightening was particularly evident in the United States and Japan, but it was also observable to a lesser extent in most Western European countries. The forces leading to recession were strengthened further. By the end of 1974 the sharp fall in aggregate demand was clearly acting to decelerate inflation in every country except the United Kingdom. But the cost, in most countries, was the worst recession in thirty years.

The Transition: 1975–77

By early 1975 most of the force of both the inflationary and demand-depressing effects of higher oil prices had already been felt. Monetary and fiscal policies, initially tightened against the inflationary consequences of the oil rise, were in the process of being reversed, and in most countries the impact of higher oil prices on domestic wages and prices was probably almost complete. Barring a further sharp increase in its real price, the future course of inflation should not be, in any substantial sense, traceable to oil.[14]

Over the next several years the major impact of the oil crisis on aggregate demand will be positive, as producers of energy, foreign and domestic, begin to spend a steadily rising proportion of their swollen receipts on purchases of goods and services. The increased aggregate demand will appear in two principal forms: growing exports to the oil-exporting countries and rising investment in the domestic energy sector. Exports to the nonoil-exporting developing countries probably will continue to decline (as projected in Table 1-4), but the positive forces operating on aggregate demand will outweigh the negative.

On the basis of the projections developed in the first part of this chap-

---

14. The United States may be an exception, as discussed on p. 28 below.

ter, the additional payments made to the oil-exporting countries by industrial countries because of the rise in oil prices probably peaked in 1974. As Table 1-5 shows, the additional oil payments (on the assumption that the real price of oil remains constant at $10 a barrel in 1974 dollars) should remain slightly below 1974 levels for the next several years. Combined with a recovery in national output from recession, this should produce a modest decline in the proportion of gross national income absorbed by incremental oil payments to the oil-exporting countries.

The near-term outlook for payments by consumers to producers of oil is quite different in the United States from the outlook in other countries. At the end of 1974, about 40 percent of U.S. consumption of crude oil remained under price control, with a price ceiling of $5.25 per barrel.[15] As a consequence the average price of crude oil in the United States was about $9.25, while the delivered price of imported oil was $12.50 and the price of uncontrolled domestic oil was only slightly lower. Throughout the first nine months of 1975—or until the time this book went to press (in mid September)—the President and the Congress had been unable to agree on whether price controls on oil and gas should be eliminated. In the meantime, the President had imposed a fee of $2 a barrel on imported oil. Whatever the outcome of the legislative battle over price controls and the related struggle over the future of import fees may be, it is likely to incorporate some net increase in the prices of domestic fossil fuels. As prices rise, additional demand-depressing and cost-push inflation effects will reappear in the United States. By now, however, the demand-depressing effects of oil price increases have been widely recognized. It is probable, therefore, that decontrol measures will be linked with proposals for tax cuts and other offsetting fiscal measures. For purposes of analyzing the transitional period, we have assumed that the demand-reducing effects of any further significant price increase for domestic crude oil in the United States will be neutralized by appropriate demand-management actions.

The rise in exports to the oil-exporting countries will be quite sharp in the next several years. While only one-sixth of the increased oil-import bill was offset by incremental exports in 1974, that fraction should grow to almost three-quarters by 1977. And this improvement is likely to be

15. The figure of 40 percent includes natural gas liquids, which are under a separate form of price control based principally on natural gas prices.

Table 1-9. *Major Oil-Related Factors Affecting Aggregate Demand as a Percentage of GNP: The Combined Impact on the United States, Western Europe, and Japan, 1974–77*

Percent; signs indicate direction of effect on aggregate demand

Year	*Additional payments for oil*[a]	*Loss of exports to nonoil developing countries*	*Additional exports to oil-exporting countries*	*Additional investment in domestic fossil fuels*
	A	B	C	D
1974	−2.2	*	+0.3	...[b]
1975	−2.0	−0.2	+0.6	...[b]
1976	−1.9	−0.3	+0.9	...[b]
1977	−1.8	−0.3	+1.1	+0.4 to +0.6

Sources: GNP: 1974 actual, plus changes in 1975, 1976, 1977 assumed by Yager and Steinberg in Table 7-4. Column A, additional oil import bill from Table 1-5; additional payments to domestic producers of oil in 1974 from Perry, Table 2-4; those for 1975–77 extrapolated on the basis of movements in domestic production shown in Table 7-9. Absolute values in Columns B and C from Table 1-5. Column D, see text.
* Less than 0.05 percent.
a. Includes, in the United States, additional payments to domestic oil producers.
b. Not estimated.

offset only modestly by the continuing loss of exports to oil-importing developing countries (see Table 1-9).

It would also seem logical to expect a stimulus to aggregate demand from rising investment in domestic energy industries in the United States and Western Europe. In considering the magnitude of this increase it is useful to discuss separately investment prospects for fossil fuels (oil, gas, and coal) and for electricity.[16]

Except for the large discoveries of oil and gas in Alaska and the North Sea and of gas in the Netherlands, exploitation of domestic fossil fuels in the United States and Western Europe had been declining for some time before the October 1973 embargo; with the world price of oil falling in real terms, the cost of expanding production of oil, gas, and coal in the United States and of coal in Western Europe exceeded the cost of meeting growing energy requirements through oil imports. But with the virtual quadrupling of world prices, it has become profitable to invest in those once-declining domestic sources of energy. Since the development and

16. During the period of time with which we are concerned, expenditures for the development of more exotic energy sources—coal gasification and liquefaction, shale liquefaction, and solar and geothermal power—will be too small to play a significant role in the overall OECD economic picture.

production of fossil fuels is highly capital-intensive, the rise in world oil prices should eventually result in a significant increase of investment in fossil fuels in Western Europe and the United States. Japan has few fossil-fuel resources to exploit.

Investment expenditures for oil-and-gas development and coal mining have already begun to increase. Particularly in the United States, however, environmental concerns are likely to play a major role in determining the pace at which large additional reserves of coal will be exploited. The most important new sources of oil in the United States lie in Alaska and on the continental shelves, and here also environmental concerns have left long-term plans for exploration and development uncertain. In both the United States and Western Europe temporary shortages of oil-well-drilling equipment, specialized steel tubing, and the like have further hampered the expansion of investment. As a consequence, it is difficult to estimate the rate of increase of investment in the fossil-fuel sectors. Various estimates have been made, however, by official organizations and private analysts. In most cases these projections show only total investment in the fossil-fuel sectors of the economy; they do not provide a measure of incremental investment—that is, the additional investment, over and above what would have been made at 1973 prices, attributable to the recent rise in energy prices. To approximate such an estimate, we have projected investment under conditions of stable 1973 energy prices and compared it with a number of recent energy-investment projections under present conditions. The results are shown in Table 1-10.

Most of the investment projections are given only as cumulative requirements over a number of years (usually 1974–85), with no indication as to the likely year-to-year rate of change. The two estimates for 1977 in Table 1-10 correspond, first, to an assumption that incremental investment increases at a constant average rate throughout the period 1974–85, and alternatively, to an assumption that investment rises sharply by 1977 to the average level for the decade and then remains at that level.

Only for the United States do we have a number of estimates. While the range is fairly wide, it is clear that even the highest estimate ($14.5 billion a year) is not very large in relation to GNP. A part of this estimated investment, moreover, consists of advance payments to the U.S. government by private companies for the right to explore and develop areas of the outer continental shelf. These expenditures do not represent demand

Table 1-10. *Alternative Estimates for Additional Investment in Fossil Fuels Attributable to Increased Oil Prices, OECD Countries, 1977*

Billions of 1974 dollars

Country or region and source of projection	Assumption 1	Assumption 2
Western Europe		
OECD	4.0	6.0
United States		
Arthur D. Little	5.0	9.0
Bosworth and Duesenberry	7.0	11.0
OECD	7.0	11.0
Hass, Mitchell, and Stone	9.5	14.5
Japan		
OECD	0.3	0.5

Sources: OECD, *Energy Prospects to 1985*, Vol. 1, Table 3, p. 176; Barry Bosworth and James Duesenberry, *Capital Needs in the Seventies* (Brookings Institution, 1975); Jerome E. Hass, Edward J. Mitchell, and Bernell K. Stone, *Financing the Energy Industry* (Ballinger Publishing Company, 1975), Table 6-1, p. 104; Arthur D. Little, Inc., "Energy Capital Outlay as an Economic Burden, 1974–1990," A Report to the Federal Energy Administration (1974; processed). In all cases, expenditures were converted to 1974 dollars. Investment was estimated by the OECD at both pre-embargo and current oil prices. In all other instances, "base" investment outlays were estimated by extrapolating from past trends.

Our first assumption (Assumption 1) is that incremental investment increases at a constant rate throughout the period 1974–85; our alternative assumption (Assumption 2) is that incremental investment rises sharply to the average level for the decade and then remains at that level.

for investment goods but transfers of economic rents from private companies to the government.[17]

Replacement and expansion of electric utility plants account for more than half of investment in the energy sector at present. The electric-utility industry is both large and highly capital-intensive. But the impact of higher oil prices on investment in electric utilities is very uncertain. Several influences work in opposite directions. The price of oil used for direct heating has risen more than the price of electricity has, since fuel costs are only part of the price of electricity and since both coal and nuclear power can replace oil in the generation of electric power. Some substitution of electricity for the direct use of oil in heating may therefore occur. In addition, nuclear power plants are far more capital-intensive than plants fired

17. The estimates of Arthur D. Little, Inc., and of Barry Bosworth and James Duesenberry for the decade include about $2 billion to $3 billion a year of these advance payments ("lease bonuses"), but it is impossible to tell how much of this represents an addition to what would have been offered at the pre-embargo price.

by fossil fuels. Substitution of nuclear for fossil-fuel plants raises the investment required per unit of electricity expansion. On the other hand, because of the increase in the price of electricity, the historically rapid growth in consumption of electricity may be slowed in uses where it is not directly competitive with heating oil. The size of the investments required would be reduced on this account. Most projections of investment in electricity indicate that these factors offset one another approximately, leaving the growth of investment in electricity unchanged.[18] The OECD estimates for the industrial countries as a whole show a modest addition to investment in electricity as a result of higher oil prices. But most of the estimates for the United States, shown in Table 1-10, imply no acceleration of investment in electricity from pre-1973 trends—that is, the increase in total energy investment is accounted for by additional investment in the production of fossil fuels.[19]

Nevertheless, a decline in utility investment, below trends previously expected, is a distinct possibility. Should this come about, total energy investment in the United States may not rise much above pre-embargo trends, since a shortfall in investment by electric utilities offsets the rise in outlays for the development of fossil fuels.[20]

Our conclusions concerning the course of the transitional adjustments to higher oil prices can be summarized as follows: While the depressing influence of higher oil prices on aggregate demand will continue during the next several years, its magnitude will grow steadily smaller. The improvement arises principally from the large, but necessarily delayed, response of the oil-exporting countries in increasing their imports to the point that they are commensurate with their higher oil revenues. Increased investment in the energy sector is likely to contribute also, but the magnitude

18. The future course of investment in electricity hinges principally on the price elasticity of demand for electricity. Those who expect a slowdown in the growth have estimated a long-run price elasticity of 1.0 or more, thereby assigning great importance to the path of relative prices. The industry itself apparently believes that the elasticity is much lower and that basic trends in industrial technology and consumer preferences, rather than relative prices, account for the historical growth rates.

19. Indeed, the rapid historical rate of growth of investment in electricity might have been expected to fall even without the oil crisis. It would have been difficult to maintain the steady decline in the relative price of electricity, which was one of the factors behind the above-average growth rate of investment in public utilities.

20. As Perry reports in Chapter 2, this happened in 1974. However, the recession and special financing problems of utilities, rather than a downward adjustment of long-term demand expectations, may have been the reason.

of that contribution is highly uncertain, especially in the United States. And just as the internal multipliers and interactions of trade among the industrial countries magnified the demand-depressing effects of higher oil prices, so should they magnify the impact of the offsetting forces that are now in prospect.

The Developing Countries

Higher oil prices have had a sharp and singularly diverse impact on the non-Communist developing countries, ranging from the huge bonanza received by the major exporters of oil, through the smaller but still substantial windfall received by other oil exporters, to the heavy burden imposed on the largest number of countries in the group, the oil importers. In contrast to the assessment for the industrial countries—where the initial shock to aggregate demand and its subsequent repercussions are the crucial determinant—changes in the availability of foreign exchange are crucial to any estimate of the effect of higher oil prices on the developing countries. In assessing what happened to these countries in 1974 and what might happen over the rest of the decade, Wouter Tims, in Chapter 5, treats changes in the availability of foreign exchange—resulting either directly or indirectly from higher oil prices—as the key determinant of changes in the capacity of these countries to import, to invest, and to generate economic growth.

The direct impact of higher oil prices on the foreign-exchange position of the different groups of developing countries as estimated for 1974 and projected for 1977 is shown in Table 1-11. For the oil-exporting countries as a group (about twenty-five in number, accounting for perhaps one-fifth of the total population of the developing world), incremental oil revenues in 1974 ranged from multiples of the 1973 GNP of some countries to comparatively modest amounts. For the other developing countries, we estimate that the additional cost of imported oil, at $10 a barrel, over what would have been spent, at $2.75 a barrel, was $10 billion in 1974, or about 1.7 percent of their combined GNP, and that it will remain near that level for a number of years. Furthermore, higher oil prices, as they were passed through into manufacturing costs, increased the prices of other products that these countries import by a significant but undetermined amount.

To these figures must be added the effect on the developing countries

Table 1-11. *Direct Impact of Higher Oil Prices on the Developing Countries: Estimates for 1974 and Projections for 1977*

Category of developing country	Estimated population in 1973[a] (in millions)	*Incremental revenues from oil exports (+) or incremental costs of oil imports (−)[b]*			
		1974		*1977*	
		In billions of 1974 dollars	As percentage of GNP[c]	In billions of 1974 dollars	As percentage of GNP projected for 1977[d]
Major oil exporters					
High-income	10	+30.8	+154.0	+20.4	+58.3
Medium-income	64	+24.7	+41.9	+25.3	+27.3
Low-income	184	+7.7	+33.5	+10.9	+34.8
Smaller oil exporters	149	+4.0	+4.3	+6.4	+5.3
Oil-importing countries	1,797	−10.1	−1.7	−10.4	−1.5

a. U.S. Department of State, Publication 8796, *The Planetary Product in 1973* (1974).
b. Calculated from Tables 1-1 and 1-2 and estimated shares of the oil market of individual countries given in footnotes to Table 1-3.
c. Calculated from *The Planetary Product in 1973* and the *World Bank Atlas* (1974).
d. GNP for 1977 is projected on the basis of annual rates of real economic growth for the period 1973–77 as follows: for the major oil-exporting countries—high income, 15 percent; medium income, 12 percent; low income, 8 percent; for the smaller oil-exporting countries—8 percent; for the oil-importing countries—4 percent.

of the oil-induced portion of the recession in the industrial countries. As noted above (Table 1-8) the authors of the regional chapters estimate that higher oil prices reduced GNP by 2.5 percent in the United States, 2.7 percent in Western Europe, and 4.2 percent in Japan. This slowdown in economic activity, after a time lag, works to reduce export prices and volumes in the developing countries, restricting in turn their capacity to import and to invest. At its peak, the depressing impact of these factors on the GNP of the oil-importing developing countries is likely to be somewhat larger than that exerted by higher oil-import costs.[21] (For most of the oil-exporting countries, incremental oil revenues are more than sufficient to maintain, or greatly expand, import and development programs despite the adverse effects of the recession in the industrial countries.) As recovery proceeds in the OECD countries and output approaches the levels that would have been reached with normal economic growth, this effect

21. In the World Bank's econometric model a change of 1 percent in the GNP of the OECD countries is eventually associated with a change of 0.85 percent in the same direction in the GNP of the developing countries. Capital flows are assumed to be constant.

will diminish and eventually disappear, or virtually so. The depressing effect of higher oil prices on import costs, however, will continue much longer, because these countries have more limited opportunities than the industrial countries to reduce consumption of energy and increase domestic production of primary sources of energy at comparatively low economic cost.

What can be said about the adjustments that must be made by the developing countries over the next few years, assuming that the price of oil remains at $10 a barrel? Here again it is useful to distinguish the widely differing positions of the three groups of developing countries— the major exporters of oil, the smaller exporters of oil, and the nonoil developing countries.

The major oil-exporting countries, as is evident from the projections in Table 1-11, will continue to enjoy boom conditions for many years ahead, with incremental oil revenues still amounting to between one-fourth and one-half of projected GNP in 1977. In general, the oil-market shares of the low-income countries in the group should increase, as the oil exporters with higher income assume most if not all of the responsibility for restricting oil exports in order to maintain the assumed high price of oil. The incremental oil revenues of these low-income countries, therefore, which have large populations and both the need and the capacity to spend these revenues quickly, should continue to increase, while those of other major exporters of oil should remain the same or decline.

For the smaller oil-exporting countries, higher oil prices are a positive economic factor of growing significance. Incremental oil revenues of such countries as Mexico, Ecuador, Colombia, Egypt, Syria, and Angola were already appreciable in 1974, and if production possibilities were to materialize, they could become much larger, since these countries are not likely to restrict exports themselves but rather to rely on the large oil exporters to accept responsibility for maintaining high oil prices. For this group of countries we have projected incremental oil revenues at $6 billion a year in 1977, or about 5 percent of their combined GNP. To some extent these gains will be offset by losses in foreign exchange resulting from recession-induced reductions in exports to the industrial countries and by a deterioration in their terms of trade for commodities other than oil, but in almost all cases the positive effect of incremental oil revenues is likely to dominate.

For the nonoil developing countries, on the other hand, higher oil prices constitute a heavy burden and a potentially serious threat to economic expansion. These countries contain four-fifths of the total population of the developing world. The increase in the cost of their oil imports in 1974—some $10 billion—was somewhat more than the total net flow of concessional capital to them in 1973. In 1974, however, and probably in 1975 as well, the immediate impact of these higher oil-import costs has been and will be offset in part by increased receipts of capital from the members of OPEC, receipts that are estimated to average $6 billion a year in these two years, of which about half is on concessional terms. Over the longer term, capital flows are less certain, and in any event, increased burdens of indebtedness are being accumulated which will be a charge on the future availability of foreign exchange.

To analyze the prospective economic position of these countries for the rest of this decade, Wouter Tims, in Chapter 5, uses a technique that takes into account not only the increase in the cost of their oil imports but also all other factors (whether oil-induced or not) that may affect their foreign-exchange position—the volume of their exports, their terms of trade, and their prospective capital inflows. His results are thus attributable to more than higher oil prices alone, but oil looms very large in the final effect.

If these factors are taken into account, it becomes apparent that higher oil prices, for special reasons, did not reduce output appreciably in 1974, but they will have a growing adverse effect in subsequent years. In 1974, the increased cost of oil did indeed add 13 percent to the import bill of these countries, and it accounted for at least half the deterioration of $20 billion in their combined current-account balances. For three reasons these huge losses caused less of a decline in output than might have been expected. First, about half the deterioration in the current-account balance represented the elimination of unusual gains in foreign-exchange reserves realized by these countries in 1973 because of the sharp increase in prices of primary products and the booming markets for their other exports. (Financial reserves of the oil-importing developing countries had increased by $10 billion in 1973.) To this extent, the rise in the foreign-exchange cost of oil imports did not require a reduction from 1973 levels in other imports. Second, the explosion in sugar prices brought net windfall earnings to developing countries amounting to about $5 billion in 1974. Third, capital inflows in 1974 went up by $8 billion. The sources of this additional capital were oil credits and concessional loans from the members of OPEC, borrowing from the international agencies (including funds drawn from the special oil facility of the International Monetary Fund [IMF]), and borrowing from private capital markets.

For 1975 and beyond the prospects are distinctly bleaker. The drain of high import costs will go on, but in addition these countries as a group could suffer even larger foreign-exchange losses from a weakening of export markets and a deterioration in the terms of trade resulting from the recession in the industrial countries, part of which, as noted earlier, is oil-induced, and from such special factors as the collapse of sugar prices. Very large capital inflows would be needed to offset these losses. Thus, Tims calculates that even if there is a fairly rapid recovery from the recession, the oil-importing developing countries, in order to maintain economic growth at an average rate of 6 percent a year, would have to receive much more capital from abroad than now seems to be in prospect. He suggests that by 1980 the annual shortfall might be on the order of $17 billion to $20 billion (1974 dollars), or about 50 percent of the total net capital inflow to the developing countries in 1974.[22]

Obviously, this capital gap could be reduced by internal adjustments in the developing countries, such as specific measures to promote domestic saving, import substitution, or export expansion. Or, the capital shortfall might not be so large as projected, because the members of OPEC and the OECD together managed to increase the flow of concessional capital to the developing countries substantially above present levels or because the more rapidly growing developing countries are able to borrow larger amounts in private markets than were projected, or because the terms of trade may prove to be somewhat more favorable than Tims's projections suggest. Furthermore, the effects could be overstated because of the use of excessively rigorous assumptions about the role of foreign exchange in determining the level of economic growth. In all these respects, however, the uncertainties apply to the magnitude—not to the direction—of existing trends. Present prospects clearly indicate that over the next few years at least the average rate of economic growth in the nonoil developing countries will be significantly reduced from the level that had been expected before the October 1973 increase in oil prices.

Most notably, however, this analysis highlights the critical distinction among the nonoil developing countries between the rapidly growing, export-oriented countries and the more slowly growing countries, whose

22. These figures probably represent an overstatement of requirements because the additional oil revenues received by the smaller oil-exporting countries are not taken into account. See methodology in Chapter 5.

economies are restrained by a highly unfavorable ratio of population to resources.

Countries in the first group will pay three-fourths of the increase in the oil import bill of all developing countries, which is a large burden for them by any standard and one that is likely to continue for some years. These countries may be able to bear this burden more readily because of their potential for rapid economic growth. For them, the critical factor will be the rate of economic recovery in the industrial countries. If it is rapid, they can expect an equally rapid recovery in their export earnings. In the meantime, they will have to be able to continue borrowing in private capital markets and to have increased access to governmental or quasi-governmental credit—the facilities of the IMF, the IBRD, and other international financial institutions and bilateral loans from members of the OECD and OPEC. The former is probably dependent on the latter, since increased loans from governments and international financial institutions, particularly long-term loans, would help to maintain the credit standing of these countries in private capital markets. With adequate financing these countries should be able to weather the transition to world economic recovery—although at somewhat lower than accustomed rates of economic growth—and be good credit risks at the same time. Otherwise, it could be a close question.

On the other hand, the low-income developing countries, which face a continuing burden amounting to perhaps $2 billion to $3 billion a year as a result of higher oil prices alone, are in the most difficult position. They cannot borrow in private capital markets, they are unlikely to be recipients of significant OPEC investments, and their internal possibilities for adjusting to higher oil-import costs are extremely limited. They must therefore rely on the receipt of increased concessional capital from members of the OECD and OPEC to offset the increased direct cost of oil imports and the loss of foreign-exchange earnings caused by the recession. Without such additional assistance they will have to bear the burden of higher oil prices in the form of disproportionately large costs in income, consumption, and employment.

The International Financial System

Most of the initial fears about the consequences of higher oil prices centered on the system of international payments and trade, rather than on the problems of managing aggregate demand in the industrial countries. For the oil-exporting countries the increase in oil prices led to a current surplus of approximately $60 billion in 1974. For the industrial countries the corresponding phenomenon was an unaccustomed and very large current deficit. The size, speed, and special characteristics of these changes were without precedent. Perhaps understandably, therefore, they created a widespread sense of alarm—that some industrial countries would become insolvent, that oil-exporting countries would soon own a large and steadily growing portion of the world's productive assets, and that the international economic system would break down under the pressure of these financial flows.

In Chapter 6, John Williamson explains why these fears were either misplaced or grossly exaggerated. Deliberately taking the pessimistic assumption that the current surplus of the oil-exporting countries would remain at $60 billion a year, he shows that as long as the real rate of growth of the world economy held reasonably close to trend levels, the proportion of the world's capital stock accumulated by the oil-exporting countries would neither grow indefinitely nor become intolerably large.

As to possible failures in the mechanism by means of which the surpluses of oil-exporting countries are invested in the industrial countries and recycled among them, Williamson's sanguine assessment has been borne out by events. Neither a global shortage of liquidity nor a weakening, to say nothing of a breakdown, of private lending facilities has occurred. Problems arising from the handling of oil surpluses did not produce a run on the private markets. And it seems likely from the experience of 1974 that the combination of flexible exchange rates and central bank co-operation would be adequate to protect against the destabilizing consequences of any purposeful "sloshing around" of oil money, even in the improbable circumstance that Arab countries would choose to use their financial surpluses for disruptive purposes. At times, preferences by members of OPEC to hold more of their surpluses in German marks and Swiss francs rather than dollars contributed to the strength of the former currencies and the weakness of the dollar. These shifts in preference, however, probably stemmed from interest-rate movements and judgments about the future rates of inflation in the reserve-currency countries, rather than from political factors.

Indeed, private markets were responsible for recycling most of the huge flow of oil money in 1974 and have continued to perform this function smoothly until now. Concern that strains arising from oil money would be heavier than these markets could sustain slowed their growth in 1974. Now that it has been demonstrated that private markets can absorb these pressures and that the key monetary authorities are prepared to provide support to prudently managed banks as it is needed, flows of private capital are once again growing rapidly.

As a result, creditworthy countries have been able to borrow to finance their oil deficits, and international reserves have risen as a consequence. Thus by May 1975, the international reserves of the oil-exporting countries had reached $54 billion, having gained almost $45 billion since the end of 1973, while global reserves increased even more. So far, oil deficits as such have not brought about an erosion of the reserve position of most oil-importing countries. A threatened outbreak of competitive trade and payments policies, moreover, did not materialize. Presumably the worst did not happen in this instance because the recycling mechanism worked effectively and because it came to be widely understood that such actions would be self-defeating for any country initiating them and would increase the collective deficit as well. This understanding was reflected in the pledges taken by the industrial countries, both in the OECD and in the IMF, in 1974 and renewed at the OECD ministerial meeting in 1975. In effect, each country agreed that in managing the pressures of higher oil prices in its international payments it would refrain from actions that would merely pass on the burden of adjustment to other countries.

Can a similarly tolerable outcome for the international financial system be expected over the next few years as well? Specifically, will the financial flows arising from oil continue to be smoothly offset by capital transactions? In any event, will it be necessary to reach international agreement on methods of distributing oil deficits, not only to assure mutually consistent objectives among the industrial countries with respect to the balance of payments, but also to maximize the growth of the world economy by encouraging the flow of OPEC savings to those countries that are likely to realize the highest returns on their use?

Answers to these questions depend in part at least on the length of time it will take for the financial surpluses to disappear—that is, for the oil-exporting countries to be able to accept full payment for their oil in imported goods and services. If this happens fairly quickly, the problems of international adjustment will by definition diminish and disappear as they give way to the underlying domestic problems associated with the transfer of increased real resources to the oil-exporting countries.

We have argued in earlier sections of this chapter that as far as the industrial countries are concerned this transitional period should be relatively brief. At present prices, the oil imports of these countries are not likely to increase and may decline moderately, while their exports to the oil-producing countries will continue to expand rapidly. Thus, as an illustration, our projections in Table 1-5 indicate that the "oil deficits" of the industrial countries—that is, the difference between the incremental cost of their oil imports at $10 a barrel and their incremental exports of goods and services to the oil-producing countries—will decline from approximately $50 billion in 1974 to $15 billion in 1977. Carrying this projection further indicates that equilibrium could be reached soon thereafter. At that point the counterpart of the financial surplus of the oil-exporting countries will be the oil deficit of the developing countries. Furthermore, the accumulated oil deficits of the developing countries, either owed by them directly or in effect guaranteed by the OECD countries (through their responsibility for the repayment of funds borrowed by the IMF and the IBRD), will constitute a significant portion of the accumulated OPEC surplus. Obviously other flows will by then be equally or more significant. Nonetheless, this trend highlights the need, outlined earlier, to concentrate more attention on the interim financing problems of the developing countries. Through some politically feasible method of sharing the burden between the countries of the OECD and OPEC, the recycling mechanism, broadly understood, must provide a continuing means of transferring a growing portion of the diminishing annual surplus of the members of OPEC to the developing countries.

Our projections suggest that by 1980 the accumulated surplus of the oil-exporting countries as a group, taking into account interest payments during the intervening years, might be approximately $150 billion (in 1974 dollars). No precision can be attached to this estimate, but the major underlying variables, notably a prospective decline in total world exports of oil during the latter years of this decade and a steady increase in the share exported by countries with a high capacity to absorb imports, point toward a rapid decline in the rate of accumulation with total net holdings reaching a peak around the end of the decade.[23] In any event, it would be

23. This figure is arrived at by carrying forward the projections for the period 1974–77 shown in Table 1-3, using the methodology explained in the text and in the footnotes to that table. We project world oil exports to be 29.2 MBD in 1978, 28.7 MBD in 1979, and 28.0 MBD in 1980. Of this total, countries spending incremental

well within the capacity of the world economy to employ productively an accumulated surplus of this magnitude, or even one in the upper end of the range of currently plausible estimates,[24] and to absorb it without political strain. By 1980, the combined GNP and the total capital stock of the OECD countries might each be on the order of $5 trillion (in 1974 dollars), and their annual net capital formation might be $600 billion.

There is the additional question of whether OECD countries will so differ in their capacity to gain shares of the incremental market in the oil-exporting countries as to cause unusual problems in the adjustments required to achieve equilibrium among themselves. Thus far such a disparity has not been evident to any significant degree, both because all the industrial countries have been concentrating on expanding their exports to these markets and because the markets themselves are so dispersed, ranging from Indonesia to Saudi Arabia, among the major oil exporters, but including the USSR and China as well as many other countries among the smaller exporters of oil. There is thus a strong possibility that incremental exports will be well dispersed among the industrial countries, even though there is no reason to suppose that these countries will increase these exports in proportion to their incremental oil-import costs. More fundamentally, as the process of adjustment continues, the normal growth of world trade will become much larger than the incremental cost of oil imports, so that in fairly rapid order, the general factors determining payments equilibrium—notably differing rates of inflation, employment, and growth of productivity—will once again become dominant.

---

revenues over a three-year period are projected to export 9.1 MBD in 1978, 9.5 MBD in 1979, and 9.6 MBD in 1980. Countries spending incremental revenues over a five-year period are projected to export 9.5 MBD in 1978, 9.0 MBD in 1979, and 8.8 MBD in 1980. The financial-surplus countries (those having low absorptive capacities) are projected to export 10.6 MBD in 1978, 10.2 MBD in 1979, and 9.6 MBD in 1980. Economic and military grants from all oil-exporting countries combined are assumed to be $5 billion in 1974–75, $4 billion in 1976–78, and $3 billion in 1979–80. Earnings on the accumulated surplus are calculated at a rate of 3 percent a year in real terms.

On this basis, the potential financial accumulation arising out of incremental oil revenues would be $155 billion in 1980, of which the financial-surplus countries would account for $100 billion. Of the remaining $55 billion, a large proportion would consist of highly liquid investments needed to pay for imports of equipment and services that had been ordered but not yet delivered. (All figures are in 1974 dollars.)

24. For example, the World Bank has recently estimated that the OPEC financial accumulation in 1980 will be $460 billion in current dollars (*New York Times,* April 20, 1975). In 1974 dollars, this figure would be close to $300 billion.

If the adjustment process takes any such course as this, there will be little reason to question the capacity of the international financial system to accommodate the resulting shifts in international financial flows and investment. Intergovernmental financial backstop facilities are already in place, and the annual financial flows, as well as the accumulated financial surplus, will not be large in comparison to the huge size of the world economy, the likelihood of its continued growth, and the annual investment resources it will require and generate. Nor does it seem necessary to worry too much about where OPEC savings are employed, or otherwise to seek explicit international agreement on how to distribute the oil deficit.[25] In effect, a relatively brief transition from financial surpluses to the real transfer of resources would mean the rapid disappearance of any need to distinguish between oil and other factors in considering what balance-of-payments aims or exchange rate or other adjustment policies are appropriate.

Long-Run Costs of Higher Oil Prices: The Final Phase of Adjustment

As we noted earlier, the immediate economic losses from higher oil prices, at least in the industrial countries, have arisen because of lower national employment and output. Producers of energy levied a "tax" upon consumers but did not spend the proceeds. Importing nations paid much of the increased oil bill not with higher exports to OPEC but with IOUs. Aggregate demand fell; jobs lost in the consumer-goods industries were not offset by new jobs created in industries producing goods for export or investment.

In the long run the higher prices of energy products will be paid by a real transfer of resources—by exports of goods and services to the oil-producing countries and by larger amounts of capital and labor devoted to the exploration and development of lower-grade domestic fuels. While there is no reason to believe that the ultimate equilibrium will be one in which

25. Williamson, in Chapter 6, takes a somewhat different view, but only because he assumes that the oil-exporting countries will have large financial surpluses for a long time. On that assumption, he suggests that oil deficits should be allocated among industrial countries on the basis of GNP, which he uses as a proxy in indicating where OPEC savings could be most productively employed.

the higher revenues from energy price increases will be matched dollar for dollar with higher spending by producers of energy, it is convenient to analyze the final phase of the process as if that were happening and then make whatever modifications seem necessary because of a failure of the two flows to match. We assume, therefore, that in the final stage of the adjustment process the demand-depressing and cost-push inflation effects of the oil price increase will have disappeared. Standards of living will be lower than they might have been otherwise, not because of lower aggregate demand and employment but because each unit of energy consumed will use more resources than when oil was priced at $2.75 a barrel—resources devoted to producing exports to pay for imported oil and resources devoted to exploiting domestic supplies of energy.

The economic losses ultimately imposed on importing nations by a rise in the world price of oil from $2.75 to $10 a barrel will consist of a number of elements: the higher cost of imported oil, paid for by increased exports; the additional cost in resources incurred in expanding production of domestic sources of energy as substitutes for imported oil; and the losses imposed on consumers and business firms as they reduce consumption of energy in response to the higher prices. Paradoxically, it is by incurring costs of the last two types—which arise in the process of increasing the supplies of energy and decreasing the demand for energy in the importing countries—that overall economic losses are minimized. Expanding domestic supplies of energy, whose costs are higher than the pre-embargo price of oil but less than current import prices, imposes costs upon a nation, but those added costs are less than the cost of imports. The overall economic loss is therefore reduced. Similarly, if consumers and business firms, faced by higher prices, decide to switch to less energy-intensive products or methods of production, they incur costs in terms of satisfaction forgone or a loss of efficiency in production. But again, their decisions to make the switch imply that the costs of doing so are less than the costs of paying the sharply increased price for an unchanged level of energy consumption.

In estimating these losses, we will first assume that reductions in the consumption of energy or increases in domestic production of primary sources of energy do not affect the world price of oil. Obviously this will not be the case, since changes in the domestic energy situation will either reduce or increase the volume of world oil imports from the level it would have reached otherwise. A reduction in world oil imports will make it more difficult for the oil exporters, acting as a cartel, to maintain the price of $10 a barrel or to increase it, and it might force a reduction in the price. An increase in the volume of world imports would have the opposite effect on the price-setting capabilities of the oil exporters. As our second approximation, therefore, we take into account the possible effects of an alternative domestic energy situation on the world price of oil, and hence on the economic costs of each.

Estimating the Losses with Oil at $10 a Barrel

The greater the fall in consumption of oil in response to a price increase—that is, the higher the price elasticity of demand—the less the economic loss from higher oil prices. A high price elasticity would indicate that business firms and consumers could purchase substantially less energy or substitute other fuels for oil in a large way at only a modest loss of productive efficiency or consumer satisfaction. As a consequence, they would pay the higher price on a substantially reduced volume of oil and would sacrifice little in the way of consumer satisfaction or industrial efficiency because of the lower consumption of oil. A low price elasticity of demand would reflect the opposite situation: cutting back on consumption of oil would cause major losses in consumer satisfaction and productive efficiency. A large rise in price, therefore, would induce only a small decrease in consumption. Buyers would pay the higher prices on a substantially unchanged volume, and whatever reduction there was would lead to a large sacrifice of consumer satisfaction or efficiency of production.

The elasticity of domestic energy supplies—oil, gas, coal, and nuclear power—plays a similar role in the determination of the economic losses consequent upon a rise in the price of imported oil. If large increases in supplies of domestic oil, or in other sources of energy, can be secured at little increase in extraction costs, then a nation can substitute relatively cheap domestic fuels for expensive imported oil. The added costs of increasing domestic production of energy will be modest, and the volume of imports on which the higher prices have to be paid will be sharply reduced. The converse holds true if the elasticity of domestic energy supplies is low—that is, if additional domestic supplies can be brought in only at very large increases in costs.

In the short run, say one or two years, the elasticity of both the demand and the supply of energy are quite low. Motorists will drive somewhat less because of higher gasoline prices, but only as the stock of "gas-guzzling" cars is gradually replaced by more gas-conserving models will the full effect of higher gasoline prices be felt. Machinery and equipment will be designed to use less energy, and homes built to minimize heat loss, but a nation's existing stock of houses and business assets cannot be replaced overnight. In the long run, therefore, a nation can adjust to higher energy prices with less economic loss than in the short run.

In addition to the national economic losses arising from the higher import bill, the loss of consumption, and the switch to more expensive domestic fuels, there is an internal transfer of income from consumers of energy to the profits of domestic producers of energy. This arises because the price paid for domestic production of energy will tend to equal the cost of extracting resources from the least favorable deposit. Owners of resources that are more favorably located and easier to extract will get a windfall, or economic rent. This is not an economic cost to the nation as a whole, but an internal redistribution of income. It is, nevertheless, a loss to consumers. Part of the windfall gain is captured by the government through taxes on profits. In the United States, an additional part is captured through "lease bonus" payments made by private firms to the government for the right to explore and develop oil-bearing areas on the outer continental shelf.

Estimates of the price elasticity of demand for and supply of energy are highly uncertain. Whatever statistical technique is used, the estimates must rely principally on observations of the relationship between prices on the one hand and consumption on the other during historical periods in which the price of energy varied only moderately and around a level only one-third as high as the present level. The relationship between consumption and price, when the price of oil varies between $2 and $3, may not be valid in predicting the consumption response to a price of $10. Similar considerations apply in estimating supply elasticities. Moreover, a large part of potential future oil supplies in the United States lie offshore on the outer continental shelf and in Alaska, in areas where few wells have been drilled. Predictions of available quantities of oil and the costs of recovery vary substantially from one expert to another. Finally, over the longer term, the size of economic losses depends on the potential costs of producing energy from less conventional sources, such as oil shale, coal gasification, breeder reactors, and solar power. Here the uncertainties are even greater.

The long-run economic loss from the rise in oil prices to $10 is difficult

Table 1-12. *Total Costs to Consumer, National Economic Losses, and Transfers of Income from Higher Oil Prices in the United States, Western Europe, and Japan in 1980*

Percentage of GNP

Country or region and projection	Total cost to oil consumers	National economic loss				Transfer of consumer income to domestic producers
		Total	Transfer of resources to foreign producers	Increased cost of domestic production	Loss from reduced consumption	
United States	3.3	1.3	0.7	0.2	0.4	2.0
Western Europe	4.5	2.6	2.0	0.1	0.5	1.9
Japan	4.7	4.5	4.1	...	0.4	0.2

Sources: For Western Europe and Japan, calculations are based on data from OECD, *Energy Prospects to 1985*, Tables 2-2, 2-5, 2-7, and 3-1.

For the United States, calculations are based on computer printouts furnished by the FEA underlying the summary estimates of supply, demand, and price for the "business as usual" projection presented in *Project Independence Report.* (A sample of the computer printouts for the $11 case is reproduced in Appendix A IV, Attachments III and IV, of the *Report.*) The FEA results are given only for oil import prices of $7 and $11, in 1973 dollars. The calculations of cost for the United States are based on additional estimates, made by the authors, of the supply-and-demand situation that would result in 1980 from an oil import price of $4, utilizing estimates from partial submodels appearing in various sections of the FEA Report.

to estimate, and our calculations are therefore subject to a wide range of error. We have taken two sets of official estimates of demand for and supply of energy in 1980 and have calculated from them the economic loss and the redistribution of income. In both of these estimates—one for Japan, Western Europe, and the United States made by the OECD and one for the United States made by the Federal Energy Administration (FEA)—the continuation of current world oil prices is assumed.[26] These costs and losses are summarized in Table 1-12.

The economic loss to Japan (4.5 percent of GNP in 1980), which has little opportunity to substitute domestic sources of energy for imports, is significantly higher than that of Western Europe, which in turn is larger than that of the United States. It is important to remember that these losses represent neither absolute reductions in living standards below pre-embargo levels nor reductions in the rate of growth of GNP. Rather they measure the difference in living standards between a world with an oil price of $3 and one with an oil price of $10 in 1980. In the case of Western Europe the estimated loss of 2.6 percent means that instead of growing by 30 percent over the period 1973–80, per capita living standards may

26. OECD, *Energy Prospects to 1985*, principally Tables 3-1 and 3-2; FEA, *Project Independence Report*, principally Tables P-1 and P-2.

rise by 27 percent, and in the United States the economic loss from the $10 oil price means a growth in living standards of perhaps 20 percent instead of 22 percent. The higher loss for Japan must be interpreted in the light of a rate of growth expected to be faster than that in Western Europe and the United States. Over the 1973–80 period a projected rise of 45 percent in Japanese living standards on the assumption of a price of $3 a barrel for oil may be lowered to 39 percent by the rise in the price of oil to $10 a barrel. The central fact that emerges from the calculations is that, taken in the proper context, the long-run economic costs of the rise in oil prices, while far from trivial, are seen to be still relatively small. In no sense do higher oil prices, in themselves, threaten any major diminution in the long-term growth of economic welfare.

In all three regions, higher energy prices moderate the growth in consumption of energy and reduce the economic loss generated by higher oil prices. In Japan, for example, were it not for the slower growth in consumption of energy the economic loss would equal about 5.5 percent of GNP instead of 4.5 percent. Similarly, a shift in consumption from oil to other, less expensive, forms of energy moderates the loss. In all three regions, oil consumption as a percentage of total energy use is expected to be much lower in 1980 at the present price of oil than it would have been at the pre-embargo price (see Table 1-13). This shift occurs in Japan as well, but Japan will remain more dependent on oil than the other two regions.

Finally, in Western Europe and the United States, the projections imply a sharp increase in production of domestic sources of energy—oil, coal, natural gas, and nuclear power—and a sharp reduction in dependence on imports (see Table 1-13). Since the United States and Western Europe were less dependent on imports than Japan was to begin with, their economic losses will be substantially less than those of the Japanese.

The direct cost of higher prices to consumers of energy is substantially larger than the national economic loss. Shifting from imported to domestic oil reduces the national economic loss, but consumers still pay the higher prices—to domestic producers rather than to foreign countries. A large part of the additional payments to domestic producers represents increases in profits rather than in costs of extraction, and national governments will recapture a significant fraction of these through taxes on profits.

The size of the estimated economic losses shown in Table 1-12 depend,

Table 1-13. *Shifts in Dependence upon Imports and in Consumption of Oil in the United States, Western Europe, and Japan, 1972 and 1980*
Percent

Country or region and ratio	1972 actual	1980 projections	
		1973 oil price	$10 oil price
Energy imports/total energy use			
United States	15.5	28.2	10.7
Western Europe	65.3	59.7	43.3
Japan	90.4	87.1	84.4
Oil consumption/total energy use			
United States	45.8	50.2	40.5
Western Europe	63.1	66.0	50.0
Japan	77.1	72.8	69.5

Sources: For the United States, FEA, *Project Independence Report*, and the authors' estimates; for Western Europe and Japan, OECD, *Energy Prospects to 1985*.

of course, on the specific assumptions made by the OECD and the FEA about the elasticity of demand and supply. The FEA projection for the United States, for example, implies a sharp reversal, by 1980, of several historical trends. During the 1960s oil consumption in the United States was rising as a fraction of total energy use. And, beginning in the early 1970s, domestic production of oil and gas began to fall. Imports had to grow not only to satisfy the rapidly rising consumption of oil but also to replace the decline in domestic output. Without a reversal of these trends, the ultimate economic loss would be much larger, and the growth of the world oil market would of course be similarly larger, strengthening the ability of OPEC to initiate further price increases. The FEA projections for the United States also incorporate an estimate that production of coal can be sharply expanded at only small increases in cost. This has the effect, in the long run, of substantially moderating both the national economic loss and the transfer of income to development of existing coal resources.[27]

27. In the FEA projections it is also assumed that natural gas prices in the United States will be deregulated. Once this occurs, the FEA energy model implies that natural gas prices will fall as oil prices rise. About 30 percent of U.S. natural gas is produced with oil as a joint product. Higher oil prices lead to higher associated gas production, shifting the supply curve of natural gas to the right. The elasticity of demand for gas in relation to the price of oil is relatively small in the FEA model, so the upward shift of gas supply is larger than the upward shift of gas

Table 1-14. *Alternative Estimates of Oil Imports and of Costs to Consumers, National Economic Losses, and Income Transfers from Higher Oil Prices: The United States, Western Europe, and Japan Combined, 1980*

Alter-native case	Oil imports (millions of barrels a day)	Costs and losses (billions of 1974 dollars)			Costs and losses (percentage of GNP)		
		Total cost to consumer	National economic loss	Transfer of income to do-mestic producers	Total cost to consumer	National economic loss	Transfer of income to do-mestic producers
I	40.1	194.2	133.4	60.8	4.4	3.0	1.4
II	22.5	183.2	110.6	72.6	4.1	2.5	1.6
III	15.8	177.6	100.3	77.3	3.9	2.2	1.7
IV	27.5	187.7	117.7	70.0	4.2	2.6	1.6

Source: Calculations from data cited as sources for Table 1-12.
Alternative cases:
I. Assumes zero elasticity of demand and supply—that is, no adjustments.
II. Base case from Table 1-12.
III. Assumes 25 percent *greater* elasticity of supplies of oil in the United States and 50 percent *greater* net elasticity of demand for oil in all regions compared to the base case.
IV. Assumes 25 percent *lower* elasticity of U.S. fuel supplies and 50 percent *lower* net elasticity of demand.

We have made some rough calculations of national economic losses in the three regions, using alternative assumptions about elasticities of supply and demand, in order to illustrate the importance of those assumptions in the estimates. The results of these calculations are summarized in Table 1-14. Several general observations emerge from an inspection of the differences between the various alternatives. The production of additional domestic fuels, at costs above pre-embargo levels but lower than the price of imported oil, does impose an economic cost. Similarly, conserving on the use of energy is not without cost. But these additional costs are less than the gains made possible by reducing the volume of high-priced imports.

---

demand, leading to the paradoxical conclusion that the price of gas falls when the price of oil rises. The combination of this facet of the FEA analysis with the relatively optimistic assumption about the supply price of coal tends to moderate the national economic loss substantially and to moderate even more the transfer of consumer income to domestic producers consequent upon the rise in world oil prices.

As the estimates in Table 1-14 illustrate, the effect on imports of changing assumptions concerning elasticity is very large, the effect on national economic losses is moderate, and the effect on total costs to consumers is relatively small. Higher elasticities of demand and of domestic supplies of energy, for example, lead to significant cuts in consumption of oil and in imports. The volume of imports and the import bill drop by 30 percent when the base case is compared to the higher-elasticity case. But the economic costs of reduced consumption and increased output of high-cost domestic resources offset a good part of these gains, with the result that the national economic loss drops by only 10 percent. Consumers pay increased "rent" to domestic producers, moreover, so the overall costs to consumers fall by only 3 percent. Conversely, with lower elasticities, imports increase substantially, while costs to consumers and the national economic loss increase only moderately. In sum, as long as the world price of oil is assumed to be fixed in all circumstances, the most important effect of alternative elasticity assumptions is on the volume of oil imports and therefore on the vulnerability of the oil-importing countries.

Taking Account of the Effects on the World Price of Oil

These conclusions are substantially altered when consideration is given to possible effects of changes in the level of oil imports on the world price. The smaller the size of the demand for the exports of the oil-producing countries, the less the likelihood of further increases in oil prices and the greater the probability of an actual reduction in the real price of oil. Changes in world oil prices in turn would have a major effect on the size of the economic losses.

The factors that influence the members of OPEC in setting prices and the circumstances that give the cartel cohesion include a number of elements, both economic and political. Of these the volume of oil exports is the most important. Since a strict relationship between the size of the world market for oil and its price does not exist, we cannot quantitatively estimate the consequences for world oil prices of greater demand-and-supply responses in the oil-importing countries. But a few calculations can be useful in suggesting the importance of demand-and-supply responses in reducing the size of the national economic losses resulting from higher oil prices. Three illustrative calculations are given in Table 1-15.

Table 1-15. *Consequences for Imports and National Economic Losses of a $3 Tariff Imposed by the United States, Western Europe, and Japan, Using Three Different Assumptions concerning the World Price of Oil*

Assumption	Oil imports, millions of barrels a day	National economic losses	
		Billions of 1974 dollars	Percentage of GNP
Base case	22.5	110.6	2.5
Effect of $3 tariff with:			
No change in world price of oil	19.0	113.0	2.6
$1 reduction in world price of oil	20.1	107.4	2.4
$3 reduction in world price of oil	22.5	87.3	1.9

Source: Estimated by the authors on the basis of data cited as sources for Table 1-12.

They begin with the elasticities of supply and demand used by the FEA and the OECD in the base case (Case II in Table 1-14). We then assume that a $3 tariff is simultaneously placed on imports by the United States, Western Europe, and Japan. Three different and quite arbitrary results are postulated: world oil prices are unchanged; world oil prices fall by $1 a barrel; and world oil prices fall by $3 a barrel.

If world oil prices do not fall, imposition of the tariff reduces the volume of imports by more than 15 percent. Vulnerability is thereby reduced, but there is an economic cost in that national economic losses rise somewhat. Should imports of the three major importing regions fall to 19 MBD in 1980, however, total world imports would be in the neighborhood of 25 MBD. This is to be compared to world imports of 32 MBD in 1973 and to a potential production capacity of more than 50 MBD in the exporting nations in 1980. Prices may be shaded as exporting countries try to adjust to the shrunken market.

Should prices fall by $1, the three regions would import 20 MBD— still 2.5 MBD less than the base case. The economic costs of imposing the tariff would be slightly more than offset by the gains of lower import prices. The economic loss would be a bit lower than the base case. In comparison to the base case (no tariff), some improvement in vulnerability would have been achieved, together with a small reduction in economic costs.

The third case shows the consequences should the fall in imports induced by the tariff lead to a major price break. We assume, again arbitrarily, that the world price of oil is forced down by $3 a barrel. In this case the market price of oil to domestic users would remain where it was before the tariff; the price cut just cancels the tariff. Imports and domestic energy production would therefore be the same as in the base case. Vulnerability would be unchanged, but the economic loss would be sharply reduced, to 1.9 percent of GNP from the 2.5 percent of GNP in the base case, a saving of $33 billion.

While the price assumptions used in these calculations are quite arbitrary, and the elasticity estimates very uncertain, two major conclusions do emerge:

First, the benefits to any one nation from taking unilateral measures that restrict oil imports are much smaller than the benefits to each of concerted action. For example, if the United States alone imposed the $3 tariff, it would undergo the economic costs calculated above; its own imports would fall by 1.7 MBD, with a much smaller potential effect on world oil prices than the reduction of 3.5 MBD that would be possible if all three regions imposed the restrictive actions.

Second, there is a trade-off between reducing economic vulnerability and reducing economic costs. If measures are taken to restrict imports and the world price does not fall, additional economic costs are incurred, but the volume of imports is reduced and vulnerability is less. If the restrictive measures do lead to a fall in world prices, however, substantial economic gains can be realized, but the reductions in the world price tend to cancel the effects of the tariff restrictions on the volume of imports.[28]

Effects on GNP

How will the long-term economic losses estimated in Table 1-12 show up in the measures of GNP for the importing countries?

That part of the loss attributable to the higher import bill will not lead to a reduction in the overall level of output, when it is measured, as is customary, in the constant prices of some base period. Rather it will take the form of a transfer of output from domestic uses (consumption, investment, government spending) to exports. That part of the economic cost

---

28. In these illustrative calculations we have used a tariff as the policy measure taken to restrict imports. This was solely for the sake of simplicity. In the circumstances of any particular country, other types of restrictive measures may be more desirable—an excise tax on gasoline, for example, or an import quota.

represented by the loss of satisfaction on the part of consumers who, because of higher prices, switch to less energy-intensive products will not show up at all in the statistical measures of GNP, for the GNP cannot measure satisfaction forgone.

On the other hand, losses attributable to the higher cost in resources of domestic production of energy will be reflected in lower total output. Each additional unit of energy from domestic sources that is substituted for imports will require more resources (capital, labor, and materials) than the same unit of energy would have cost if purchased abroad at pre-embargo prices. The additional resources have to be diverted from the production of other goods and services, thereby lowering total GNP. Similarly, some part of the conservation induced by higher energy prices occurs not because consumers forgo energy-intensive products but because business firms switch to less energy-intensive methods of production. In turn, everything else being equal, this switch would be expected to lower the overall productivity of the economy somewhat, thereby leading to a reduced rate of economic growth. While it is impossible to make precise estimates, the loss from these sources is not likely to be large. One recent study by the World Bank projected the effect of a $5 increase in oil prices on the growth of measured GNP in the OECD countries between 1973 and 1985. Taking account of both kinds of losses that show up in measured GNP—the higher resource costs of domestic energy production and lower growth of productivity in the nonenergy sector—the study estimates that overall economic growth would be reduced by about one-tenth to two-tenths of one percent a year.

Our measures do not take into account the inevitable transition losses incurred in the process of adjusting the economic structure to a regime of sharply higher energy prices. Even with highly successful policies of overall demand management, frictional losses are inevitable as resources are transferred to new patterns of production. Existing investments in energy-intensive machines and industrial processes that were profitable when energy prices were low suddenly become obsolete in a regime of doubled energy costs and have to be retired prematurely. As demand shifts from one kind of output to another, workers can lose the advantage of previously acquired skills, and even when they have been re-employed, they may suffer losses in productivity and earnings. In the course of time these losses diminish substantially and eventually disappear, since both capital and labor are quite mobile in the long run, but in the interim—and quite apart from shortfalls in aggregate demand—there are transition costs not captured by any of our measures.

Earnings on OPEC Financial Claims

In our calculations of losses it is assumed that ultimately the oil-importing nations transfer real resources to the oil-producing nations by way of higher exports to them, in an amount equal to the higher oil bill. But in the interim period, the transfer of real resources is postponed; the oil-producing countries accumulate financial claims against the importing countries that yield returns in the form of interest or equity earnings. Ultimately the earnings on these financial assets must themselves be paid in the form of a transfer of additional resources from oil-importing to oil-exporting nations, and this additional transfer can also be thought of as an economic cost that must be paid in the future. To express this another way, part of the current payment of higher oil prices has been postponed into the future, and as is the case with any borrowing, interest must be paid on it.

The magnitude of these interest payments is not likely to be large, however, in comparison to the size of the industrial economies. The projections of oil imports and of exports to the producing countries made earlier in this chapter suggest a total accumulation of financial claims growing to perhaps \$150 billion in 1980 (in dollars of 1974 purchasing power) and not rising much, if at all, thereafter. If we assume a real rate of interest of 3 percent,[29] the resulting flow of real interest earnings to the oil-producing countries might amount to one- or two-tenths of one percent of the combined GNP of the OECD countries.

These interest payments are not a cost, to be added to the economic losses calculated earlier. In the first place they arise because the real transfer of resources has been shifted in time—from the present to the future. Unless the interest paid to OPEC is greater than the social discount rate in the borrowing countries—which it is not likely to be—the value

29. If prices rise by, say, 5 percent per year, a nominal rate of interest of 8 percent is consistent with a real rate of 3 percent, for of the 8 percent earned by investors 5 percent is required simply to preserve the real purchasing power of their portfolio, leaving 3 percent for true net earnings. For convenience we have made our calculations in dollars of constant purchasing power and have therefore used a real interest rate. Converting the estimates of asset values and GNP to current dollars and using a nominal interest rate would have given the same results.

of the postponement is worth what is paid to postpone the payments. We do not normally think of a borrower, who pays market interest rates, as having suffered an economic loss by virtue of his decision to borrow.

In the second place, if demand-management policies in the OECD countries emphasize the stimulation of domestic investment, to make up the losses in aggregate demand during the transition period, the "borrowed" resources could be used to add to productive assets. The yield from domestic investment, in terms of added growth in productivity, is likely to be greater than the real interest rate paid to OPEC. As a consequence, the OPEC financial surpluses, borrowed at one rate and invested at a higher rate of return, can be utilized to generate a stream of additional national output sufficient not only to pay the interest costs but to yield a net return, which could help to defray the higher costs of oil itself.

Policy Conclusions

The analyses of adjustment problems undertaken in the various chapters of this book have a number of implications for policy. They are spelled out in some detail in the remaining pages of this chapter. Before elaborating on them, we think it worthwhile to underline one central message that comes through with overwhelming force: in the long run, with full employment restored and the oil-induced inflation absorbed, the cost to the industrial countries of paying $10 a barrel for oil, while far from negligible, poses no significant threat to living standards or economic growth. The required long-run adjustments are not large in view of the size of the economies of these countries and their ability to reallocate responses. Similarly, problems arising from the financial surpluses of the members of OPEC are inherently manageable and will in any event steadily diminish. And earlier fears that members of OPEC would accumulate huge and constantly growing claims against the wealth of importing nations turn out on close examination to have been greatly overstated. On the other hand, two problems arising out of the oil crisis provide real grounds for concern: the short-run difficulty of managing demand so as to avoid additional unemployment and inflation brought on by the shock of a large and sudden increase in oil prices and the tendency of higher oil import costs and an oil-induced world recession to impose disproportionately large costs on the nonoil developing countries.

From a technical standpoint even the depressing effects on aggregate demand of a large and sudden oil price increase could be handled by carefully synchronized internal policies in the industrial countries. Monetary and fiscal policy can restore the aggregate demand lost to higher oil prices, so that output and employment need not suffer to any large degree. But pursuit of optimum policies to maintain employment and output runs up against a major barrier. Under the social and institutional arrangements of modern industrial societies, the initial effect of higher oil prices in raising the overall price level will almost certainly set in motion an even larger inflation. When prices rise—even though the rise is externally imposed and must be borne—claims for wages and other factor incomes are likely to be escalated in a futile attempt to recapture the losses. In the absence of some social mechanism to determine the way in which income shares should be adjusted to the shock, policy makers, with only demand-management tools at their disposal, are faced with a dilemma: if output and employment are maintained, the external shock will generate successive rounds of inflation; if inflation is to be avoided, output and employment will have to fall.

This line of reasoning leads to two basic conclusions about the desired direction of policy: *First,* given the imperfections of internal social institutions, concerted policies adopted by the industrial countries to reduce oil imports have several advantages. By shrinking the size of the world oil market, they impair the ability of the oil exporters to exert their will on oil prices and thus to impose large shocks on the rest of the world. At the same time, they reduce the size of the economic shock associated with any given price increase (since oil imports become less important in the affected economies). *Second,* the achievement of a modified incomes policy that incorporates a social consensus on the way to manage and share the income losses is the only way to make it possible for policy makers to minimize the impact of any new external economic shock from oil on employment without risking serious inflation.

A More Detailed Review of Policy Implications

It becomes clear with the benefit of hindsight—and it should have been clear at the time—that the uncompensated loss of consumer purchasing power stemming from the OPEC price rise would generate a substantial reduction in aggregate demand in the industrial countries. But the same price increase also gave an additional jolt to an inflationary process that had already been accelerating sharply. On the assumption that prices of other goods and services were not likely to fall, the direct pass-through of higher prices of oil and other sources of energy into the prices of final products simultaneously measured the immediate inflationary effect and, to a reasonable approximation, the demand-reducing impact of the oil price increase. Since the importing nations did not immediately have to pay for the higher-priced oil with correspondingly higher exports or greater domestic energy costs, it would have been possible to aim at off-setting the reduction in demand through stimulative fiscal and monetary measures while limiting the inflationary effect to a one-shot direct pass-through of the higher fuel prices.

To limit the inflationary effect of the oil price rise it was critical to pre-vent wages from increasing in response to the initial rise in oil costs. An attempt to regain the lost purchasing power through wage increases was bound to be self-defeating, for that would simply have led to more infla-tion, via the cost-push mechanism.[30]

Since payment in real resources for most of the cost of higher oil prices was being postponed, a tax reduction could have maintained after-tax real incomes without raising wage costs, and at the same time it would have offset the demand-depressing effects of the oil price increase. To be effec-tive in moderating inflation, such a policy would have required some kind of explicit arrangement between the government and union leaders to count the tax reduction as equivalent to a wage increase. To put this an-other way, the purpose of such an agreement would have been, in effect, to remove the rise in oil prices from the cost-of-living index, insofar as changes in the cost-of-living index formed the basis for wage bargaining and adjustments.

In the absence of some such social compact, aggregate demand was allowed to fall, as the only other means of isolating the initial price increase and preventing a new wage-price spiral. Indeed, far from desiring to com-pensate for the loss in purchasing power, some welcomed the oil price increases an as "excise tax" that imposed badly needed demand restraints

---

30. In theory the money wage increase could come at the expense of profits, allowing real wages to remain unchanged despite higher oil costs. If profit margins are abnormally high to begin with, as was the case in early 1951 after the upsurge of prices in 1950, this is a possible outcome. But under more usual circumstances the wage increase will almost surely be substantially passed through into higher prices.

on an inflationary world in which political caution prevented policy makers from pursuing sufficiently spartan economic policies.[31] In most countries, therefore, not only was there no stimulative offset provided, but fiscal and monetary policy became even more restrictive in response to the increase in inflation that higher oil prices generated. As a consequence, wages did not adjust fully to the higher prices, and by late 1974 price inflation did begin to decelerate. But the cost of pursuing this path was a large increase in idle resources and, in some countries, the severest recession of the past thirty years.

Even though policy makers consciously chose to let aggregate demand fall as the means of isolating the initial inflationary shock, it is also true that they got far more decline in demand than they bargained for. The magnitude of the demand-depressing effect of higher oil prices simply was not appreciated. Fairly accurate statistical estimates were certainly possible;[32] the data were easily available, and with hindsight the calculations seem obvious. But the demand-depressing effect of the jump in oil prices was unique. Neither inflation nor changes in the terms of trade can usually be expected to generate large and immediate increases in unemployment. What was different about the oil price rise was that so much of the proceeds were not respent—either by the members of OPEC or by domestic oil producers. Given the novel character of the shock and the inflationary environment in which it occurred, it is not surprising that policy makers throughout the industrial nations paid only minimal attention to the potential recessionary consequences of the OPEC action.

A large increase in the cost of importing a major commodity or group of commodities whose demand elasticity is low will inevitably require a reduction in real wages.[33] If, as in the case of the 1974 oil price rise, the recipients of the higher prices do not spend the proceeds, the fall in before-tax real incomes can be offset temporarily by fiscal measures designed to raise after-tax incomes. The adjustment to the loss in real income can then

31. For example, Otmar Emminger, Deputy Governor of the Central Bank of the Federal Republic of Germany, in discussing the impact of higher oil prices on the world economy on 27 March 1974, emphasized his concern over the inflationary consequences and dismissed fears that it would heighten the danger of a world recession. See the *New York Times,* 28 March 1974.

32. As early as January 1974, George Perry and Walter Heller estimated a $20 billion demand-depressing influence that oil price increases had had in the United States by that time. Walter W. Heller and George L. Perry, "The U.S. Economic Outlook for 1974" (National City Bank of Minneapolis, 1974; processed).

33. But see footnote 30, p. 58.

take place gradually. If, on the other hand, the increases in cost must be paid immediately by a transfer of real resources—as, for example, in a devaluation unaccompanied by an increase in foreign borrowing—then not only must real wages fall immediately but there is no room for tax measures designed to maintain after-tax real incomes.[34] In either event, if a new wage-price spiral is to be avoided, money wages cannot be adjusted upward in an effort to match the price increase. But if management of demand is the sole policy tool available, prevention of a wage-price spiral can only be purchased at the cost of increased unemployment and therefore at a much greater loss in real income than that required by the cost increase itself.

In the kind of situations we have described a nation must adjust real wages. If it does not do so by adopting an explicit set of arrangements about income shares, including possible changes in the burden of taxes, the normal wage-and-price-setting mechanism in modern societies will seek to avoid the adjustments that are necessary. This will virtually guarantee some combination of inflation and unemployment, whose consequences in lost income will almost surely far exceed that required by the external shock. However costly in terms of output and employment forgone, demand restrictions in the industrial countries have finally brought about a sharp deceleration in inflation (except in the United Kingdom). Money wage increases have moderated. Finally, after the wringing-out process had proceeded for a year, expansionary monetary and fiscal policies were introduced in most countries.

In planning demand-management policies over the next several years, policy makers will have to take into account the continuing influence of the 1974 oil price rise. As the projections of the three chapters dealing with the United States, Western Europe, and Japan indicate, the drain of purchasing power can be expected to continue through 1977. The shortfall in employment and output, which continued into 1975 in most industrial nations, contained more than simply cyclical elements. Other things being equal, therefore, demand-management policies cannot be formulated on the assumption that, once turned around, the industrial economies can be expected to snap back all the way to full employment by the normal cyclical factors alone.[35] As pointed out earlier, however, for the industrial

34. In the short run, of course, some of the impact of a devaluation is borne through a reduction in the profits of exporters or importers or both.

35. In more formal terms, the saving function at full-employment levels of

countries taken as a whole, the oil drain should steadily decrease in magnitude, and this too should be taken into account in formulating policy.

Three other aspects of adjustment policy require elaboration:

The first point, which can be disposed of quickly, is whether special action is necessary to influence the investment choices of the members of OPEC. Our calculations suggest that the total amount of OPEC investment will not be overwhelming—perhaps $150 billion (in 1974 dollars) by 1980—and that these holdings will be concentrated among a few countries with conservative financial objectives. (Saudi Arabia and Kuwait together could account for a least half of the total). When making their financial decisions these countries are likely to stress safety of yield and diversification rather than managerial control; hence their assets probably will reflect a predominant preference for portfolio over direct investment. Furthermore, a significant portion of total OPEC financial holdings will take the form of loans to governments of oil-importing countries and to international financial institutions. In these circumstances, government actions designed to attract, to channel, or otherwise to control such investments would be unnecessary. Market forces, backed up by co-operative arrangements among the authorities of the major central banks, should be a satisfactory determinant of how these investments are distributed between long-term and short-term holdings.[36] Moreover, existing legislation governing the ownership and operation of business enterprises in the countries where these investments are made should provide sufficient protection against their being used for other than commercial purposes.

Indeed, rather than being a problem, it is conceivable that the financial surpluses accumulated by countries of OPEC could be converted by appropriate demand-management policies into a potential boon. In effect, by draining away purchasing power from consumers in the industrial countries and not respending the proceeds immediately OPEC has raised the world rate of saving. If demand-management policies in the industrial countries emphasize the expansion of investment as the means of employing the resources initially idled by this increase in savings, the rate of long-

---

income has shifted upward by an amount equal to the oil drain. This is roughly equivalent to a corresponding increase in the full-employment budgetary surplus.

36. In contrast with this general view, Giorgio Basevi, in Chapter 3, suggests the possible need for tax preferences in the OECD countries to attract investments from members of OPEC.

run economic growth can be raised.[37] Since, as we indicated earlier, the real rate of interest paid to members of OPEC on their savings is likely to be substantially less than the productivity of the additional investment that such savings could make possible, a net return in the form of higher national income could be generated in the importing countries. While the added income would by no means equal the long-run economic losses from higher oil prices, it would offset some of their effects.

A second point concerns the character of international monetary adjustments to higher oil prices. For industrial countries, the situation is likely to require a return to normal patterns much sooner than may have been anticipated. We have argued that OECD exports to the oil-producing countries will continue to increase rapidly, while the effects of higher oil prices and, for the time being, the world recession will continue to constrain world oil imports. Within a few years incremental oil payments of the OECD countries taken as a group should be largely offset by incremental exports to oil-producing countries. Moreover, as the volume of international trade resumes its rise, oil should account for a diminishing portion of total trade. As this happens, there will be less reason to make any special distinction between oil and other factors in international transactions, and less need as well for special financing facilities among OECD countries on this account. Changes in nonoil transactions will again become the dominant element in the determination of a country's payment position.

Obviously, the situation of individual OECD countries will differ widely both in the extent to which they can reduce their dependence on imported oil and in their ability to exploit the growing market opportunities in oil-producing countries. For some countries, deficits in oil-related transactions will persist, while for others such deficits will turn into surpluses. Therefore, as the total OECD deficit in oil-related transactions diminishes and eventually disappears, it will be necessary for those countries in a less favorable position with respect to oil payments to improve their current balance with the other nonoil countries, and vice versa. This is another way of saying that over the next few years the OECD countries will increasingly have to achieve balance-of-payments equilibrium by relying

37. Since consumption has fallen sharply, an investment-expansion policy must go hand in hand with some expansion of consumption. Otherwise, the effect of increased investment incentives might be thwarted by poor market prospects for the output of the newly expanded and modernized capacity.

less on borrowing or lending and more on the underlying forms of adjust-ment—demand-management policies and appreciation or depreciation of exchange rates.

The third and perhaps most important point concerns the special diffi-culties of the nonoil developing countries. For them, in contrast to the situation in the industrial countries, higher oil prices will mean an inevi-table reduction in capital formation and in the rate of economic growth. As we have stressed earlier, these countries will be forced to reduce the imports that they require for economic expansion, both because they must pay more for imported oil and because the world recession is reducing their foreign-exchange earnings. This impairment in their capacity to import will reduce their rate of economic growth and will take a long time to redress. Consequently, their welfare losses from higher oil prices are likely to be relatively larger and longer-lasting than those of the industrial countries. Furthermore, their ability to adjust to higher oil prices depends heavily on factors outside their control—the speed of recovery in the OECD countries and the volume of capital inflows from the members of both OPEC and the OECD. One of the major issues facing the world economy, therefore, is how to devise politically and economically feasible arrangements to increase the flow of capital to these countries—on quasi-market terms to the economically stronger countries and on highly con-cessional terms to the poorest countries. This will be practical only if the burden is shared in substantial measure by the members of both OPEC and the OECD.

Higher Oil Prices, Lower Oil Prices

Our analysis until now has assumed that the world price of oil will remain constant at $10 a barrel (in 1974 dollars). We believe this to be unlikely. At a price of $10 a barrel the growth in world consumption of energy should decline significantly and production of primary sources of energy in the oil-importing countries should increase. The effect of such changes on the world oil market could be profound. For example, if as a result of higher oil prices (and conservation measures) the rate of growth in world consumption of energy should decline from its trend rate of 5.5 percent a year to 4.5 percent a year, world oil imports by 1985 would be reduced by 20 MBD or 40 percent below the level they would otherwise have reached. With the world oil market under such constraints, the ability of the members of OPEC to control the price would gradually erode. In the long term, therefore, we believe that the real price of oil is likely to decline, either because the nominal price falls or because it fails to keep up with general inflation.

Over the next few years, however, price forecasts are more problematical, since the effects of higher prices on consumption of energy and on domestic production of energy fuels will take time to work themselves out. The OPEC producers could attempt to take advantage of their short-term market leverage by increasing the price, or they could decide that a reduction in the real price might be the best way to maximize their long-term market position.[38] For purposes of examining policy implications, we will arbitrarily assume that the export price of oil (in 1974 dollars) is suddenly increased by $2 a barrel and, alternatively, that it is suddenly reduced by $2 a barrel.

With an increase of $2 a barrel in the price of oil, the OECD countries would face an economic shock equal to about one-fourth that experienced in 1974. Their oil import costs would increase by $17 billion. (The developing countries would face an added foreign-exchange burden of $3.5 billion a year.) The cost of domestically produced fuels in the OECD countries would also rise, but the size of the increase is difficult to estimate. In the United States the effect of an increase in the world price of oil would depend heavily on the status of controls over domestic oil prices and on decisions about the special import fee of $2 a barrel. Imposition of price ceilings on new domestic oil and maintenance of controls on old oil could limit the price rise to imported oil. Alternatively, removal of the $2 special import fee could neutralize the effect of a $2 increase in the world price of oil. But as of late summer 1975 no decision had been reached about either of these matters.[39] In Western Europe, there would be some spillover effects on the prices of coal and natural gas. The price of North Sea oil presumably would also go up, but since the volume of this production will still be fairly small in 1975–76, the effect would not be substantial. Japan would bear the full brunt of the oil price increase, since imported oil is the predominant source of its energy.

38. Joseph A. Yager and Eleanor B. Steinberg outline the major variables and the policy options confronting the members of OPEC in Chapter 7.

39. If a $2 rise in world oil prices were matched by removal of the $2 import fee, and if at the same time price controls on oil were discontinued, the net increase in average prices of crude oil in the United States would be about one-third the size of the rise that occurred in 1974.

As in 1974, increased oil-related expenditures would offset only a portion of the new "oil excise tax" during its first year. A number of the oil-exporting countries would increase their expenditure commitments, but disbursements would lag substantially, particularly in view of the already large buildup in project spending among the oil-exporting countries. It is also possible that the increase in the world price of oil would raise long-term expectations about the price of all energy fuels in the United States and Western Europe and thus stimulate investment in exploration, development, and production. All in all, the first-year effect might be a drain of aggregate demand on the order of $13 billion to $16 billion for the OECD countries combined. If this was not offset by stimulative fiscal and monetary measures, the result, after subsequent-round effects, could be an average decline of perhaps one-half to three-quarters of one percent in the rate of growth of GNP from the level that would otherwise have been attained and a somewhat larger increase in the rate of inflation because of the additional cost-push pressures. The effect would be considerably smaller than average for the United States (assuming that energy policies prevented the OPEC price increase from affecting prices of domestic energy fuels significantly) and considerably larger for both Western Europe and Japan.

A reduction of $2 a barrel in the price of oil would have almost symmetrical effects in the opposite direction. A decrease in the cost of imported oil would be the equivalent of a tax reduction in stimulating aggregate demand in the OECD countries. (In the developing countries, the saving in foreign-exchange costs would finance additional imports and generate capital formation.) Offsets to this stimulation would be a reduction in expected exports to the members of OPEC (small during the first year because of previous OPEC spending commitments) and a reduction in energy investment in the United States and Western Europe (which also would be small during the first year because of the large backlog of existing projects). For reasons noted above, the stimulative effect would be proportionately higher in Western Europe and Japan and lower in the United States. If monetary and fiscal policy before the price change had been sufficiently stimulative, a reduction in the price of oil would call for offsetting restrictive government action. If expansionary measures had been too modest, the oil price reduction would add to the prospective rate of recovery from recession without calling for offsetting government policy action. Again, the effect of the oil price change—in this case the stimulative effect—would steadily diminish and virtually disappear over the next few years.

The point worth stressing in this more or less mechanical exercise is that if governments have learned the lesson of the oil shock of 1973–74, the world economy should be able to adjust more quickly and with less cost to subsequent sudden increases in the world price of oil. First priority should be given to offsetting the depressing effect on aggregate demand and thereby avoiding unnecessary reductions in output and employment. Once this is done—that is, with the normal rate of economic growth maintained—adjusting to the long-run welfare costs is more feasible politically and can be accomplished more gradually.

Minimizing Long-Run Costs

The long-run economic costs of higher energy prices will depend directly on the expansion of alternative supplies and curtailment of demand in the importing countries and indirectly on the influence exerted by these developments on the pricing policies of OPEC. Our efforts in this book are directed principally toward an analysis of the ways in which various national economies can best adjust to a world of higher prices. It is not a book about oil policy itself, in the sense of being an analysis of alternative measures for expanding supply or reducing demand. We have much less to say, therefore, about long-run adjustment policies than about those for the short run. Several basic points that should be considered in the determination of oil policy, however, follow from our analysis.

First, to the extent that our estimates of long-run costs in relationship to GNP in the industrial countries are approximately correct, the magnitudes of the economic losses for the United States and Western Europe— and even for Japan—are comparatively small. They imply, not an absolute reduction in living standards, but a modest reduction for a while in the accustomed rate of growth of per capita consumption. While these losses are sufficiently large to warrant the adoption of policies to reduce them, they should not be viewed as in themselves threatening the prospects for achieving normal rates of growth.

Second, for the United States especially, there is a conflict between short-run demand-management objectives and long-run cost-minimization goals. The price of domestically produced fossil fuels in the United States—oil, natural gas, and coal—is being controlled substantially below the world price of oil. As a consequence, oil is being imported at world prices for uses whose values are less than the price paid for those imports. Similarly, to an extent difficult to estimate quantitatively, some domestic fossil-fuel resources which would be exploited at resource costs less than the price of imported oil are being neglected. For both of these reasons a policy of holding domestic energy prices below the cost of imported oil increases the national economic losses from higher OPEC prices. On the other hand, allowing prices of domestic sources of energy (which in the United States account for over 80 percent of total energy consumption) to jump suddenly to parity with world oil prices could introduce major new problems of demand management—a new spurt of inflation, both direct and second-round effects, and an additional drain on consumer income. A failure to cope with these problems would bring about economic losses far greater than the misallocation of resources caused by a multiple price for oil.

A reasonable solution to this problem of conflicting objectives might be to provide for a gradual rise in controlled prices toward parity with imported oil.[40] Particular fossil-fuel deposits are exploited over a large number of years. Similarly, changes in the consumption of energy are often associated with long-term investment decisions. As a consequence it is long-term expectations about prices, rather than immediate price levels, that influence decisions concerning supply and demand. Removing uncertainties, by agreement on a policy of gradual decontrol announced in advance, would raise long-term price expectations while significantly reducing the associated short-run demand-management problems.

Third, in designing policies to minimize the long-run loss from higher energy prices, perhaps the most important element is recognition of the fact that the market for energy is worldwide. An increase in the flow of North Sea oil should ultimately help the Japanese consumer, since it displaces oil imported from the world market and decreases the price-setting capability of OPEC. An expansion of the production and use of coal in the United States will ultimately benefit Belgian consumers for the same reason. As a consequence, it is in the interest of countries, such as Japan, with small domestic energy resources to invest in the expansion of alterna-

40. Gradual decontrol of below-market prices should be accomplished by reducing the fraction of resources subject to control rather than by raising controlled prices in stages. The latter approach would generate incentives to delay investment in the expectation of higher prices in the future.

tive sources of energy in other countries, whether or not they import the energy fuels resulting from this particular investment. Similarly, it is as much in the interest of the United States to export energy fuels produced from such additional investment as to consume them at home. And to the extent that capital from industrial countries makes possible an increase in the production of energy fuels for domestic consumption in the developing countries, the former benefit as well from the reduction in demand on the world oil market. Much the same reasoning applies on the demand side. Restraint of demand in one nation benefits other oil-importing nations. For these reasons there is wide scope for mutually advantageous strategies both in the encouragement of investment and in the restraint of demand among major importing nations.

Fourth, a choice between accepting either larger economic losses or less economic and political security is inherent in whatever oil policy may be adopted. Should the OECD countries, acting together, succeed in reducing imports they would probably also succeed in bringing about a reduction in the world price of oil. If so, they would reduce losses and improve security at the same time. Lower oil prices, however, would eventually lead to increased imports. Unless offsetting actions were then taken, the reduction in economic losses would be followed by a reduction in security. For the OECD countries to have cheaper oil within acceptable security bounds, they may have to take specific measures to limit their vulnerability once world oil prices are reduced. Such measures could include maintaining larger oil stocks or standby capacity, subsidizing or protecting domestic production, or taxing domestic consumption.

Fifth, our analysis of the potential adjustments from measures that reduce consumption of energy, in comparison to those that encourage expansion of the supply of energy, suggests an additional set of policy guidelines, at least for the medium term. There appears to be more room for minimizing economic losses through conservation measures than through expansion of domestic sources of energy. For example, in the 1985 base case of the FEA projection for the United States, the reduction in national economic losses through adjustments in demand and supply came roughly two-thirds from reductions in demand and one-third from expansion of supply. While estimates of both demand and supply elasticities for energy are uncertain, the latter are particularly so. Apart from new production from the North Sea and from the north slope of Alaska, the reliability of the projected increases in U.S. production of oil and gas incorporated in the FEA and OECD projections is subject to particularly large uncertainties. While still highly tentative, the most recent official reassessments of potential supplies of oil in the United States have been substantially more pessimistic than earlier views. The pursuit of expansion of supplies remains highly desirable, even in the face of considerable uncertainty about results. But success in limiting the growth in world oil imports probably will depend heavily on measures to restrain consumption.

While the effects of price increases already experienced can be expected to be felt increasingly, it may be necessary to go beyond the natural market forces during the years immediately ahead in order to restrain the growth of consumption, especially if the more pessimistic end of the range of estimates of potential supplies should begin to seem the most likely. This would mean the use of tax penalties and subsidies to encourage the more efficient use of energy and of other governmental measures to increase the substitution of other primary sources of energy for oil. Since the effect of these measures will only be gradual, prudence alone suggests that they be adopted soon. Such actions would give the nation greater bargaining leverage with the oil producers and greater policy flexibility in general, while experience reduces present uncertainties about the future demand and supply of oil and energy.

Finally, it is worth repeating a point emphasized earlier: less reliance on imports is important not only, indeed not principally, as a means of reducing long-run economic costs. The oil shock of 1973 proved to be the major cause of the world recession, a fact that is made no less somber by the recognition that failures in policy in the OECD countries contributed to this result. Optimum policies to counteract future shocks, even if technically possible, are extremely difficult to devise and apply and surely cannot be taken for granted. The chief value of reducing reliance on imported oil, therefore, must lie in reducing the influence of the oil exporters on the internal affairs of the importing countries. It is obvious that the danger of another embargo to achieve political objectives will continue as long as dependence upon imported oil remains high. It is equally obvious that as long as OPEC is able to increase the price of oil suddenly and by large amounts, the world economy will be vulnerable to shocks to aggregate demand and thus to substantial losses in employment and output.

CHAPTER TWO

The United States

GEORGE L. PERRY
Brookings Institution

THE ESCALATION OF WORLD OIL PRICES by producing nations during 1973–74 posed severe problems for all oil-consuming nations. All have had to contend with the impact of higher oil prices on the domestic level of activity, the rate of inflation, and the balance of payments. The resulting problems have been far more severe for some nations than for others, and their success in coping with them has varied. It will be some time before the ultimate effects on individual nations and on the world economy as a whole will be known.

The United States shares all these adjustment problems with other nations and is in a relatively favorable position to solve them. But the oil crisis also poses some issues that, if not unique to the United States, at least present it with somewhat different alternatives and problems from those confronting some other nations. Three facts account for the special situation in which this country finds itself. First, the United States is not only the world's largest consumer of petroleum, it is also the world's largest producer, and it still supplies two-thirds of its own petroleum needs. Second, the United States appears to be the non-Communist world's main potential supplier of significant energy substitutes for petro-

I AM GRATEFUL for help from several persons in the preparation of this paper. Jared Enzler and Saul Hymans ran my estimates of the economic impacts of the oil crisis through their econometric models. Their estimates of the way in which these impacts affected the economy in the aggregate constitute the basis for the second half of this paper. Jan Broekhuis and Nell Hahn provided assistance with research, and Evelyn Taylor typed the manuscript.

leum in the somewhat longer run: it has enormous reserves of coal and shale oil and the capacity to generate large amounts of nuclear power. Third, the United States has a special relationship with Israel that commits it, in some ill-defined but well-established sense, to the support of that country in its conflicts with the Arab states that surround it.

These facts have many different implications for political decisions affecting both the immediate and the more distant future. Considerations of particular importance for the short run are the pricing of domestically produced oil and the maximizing of production, along with plans for meeting the possibility of a new embargo on shipments of oil from the Arab states. For the long run, the dominant issue becomes the way in which a national energy balance will be struck allowing for different prices for oil and other sources of energy. Considerations include the feasibility and desirability of achieving self-sufficiency in the production of energy, or even the possibility of the United States becoming a net exporter of energy. The substitution of alternative sources of energy becomes more feasible in the long run, and estimates of supply elasticities, in particular, become extremely speculative, as new technologies and new discoveries of oil enter into the calculation.

This chapter deals primarily with the short-run problems posed by the 1973–74 oil crisis, the way in which these problems have been handled, and their effect on the U.S. economy. While the analysis starts with the year that has passed since the oil crisis, it also looks to the period immediately ahead, during which the repercussions of higher oil prices will still be mounting and important decisions remain to be made.

Short-Run Problems, Actions, and Impacts

In the fall of 1973, the announced embargo on oil shipments from major Arab producers and the clear prospect of sharply higher import prices rocked the U.S. economy. U.S. policy had to encompass two dimensions of the problem that were not of comparable concern to other nations. First, the embargo was directed at the United States because of its pivotal role in supporting Israel in the war. U.S. planning, moreover, had to contend originally with uncertainty about the effectiveness of the embargo and eventually with uncertainty about the date of its termination. Second, because the United States produces much of its own oil whereas

Western Europe and Japan do not, a price policy had to be established for domestically produced oil; this policy would be the main factor in determining the average price of oil products in the United States, since about two-thirds of domestic demand was still satisfied by domestic production.

Three distinct issues confronted U.S. policymakers:

• *The pricing issue:* How domestic products and crude oil production would be priced, taxed, and encouraged.

• *The allocation issue:* How a restricted total supply of oil, and hence of petroleum products, would be allocated among users and industries and how demand would be constrained to the available supply.

• *The macroeconomic issue:* How the macroeconomic effect of the oil crisis would be handled with respect to output, employment, inflation, incomes, and the balance of payments.

These issues are clearly interdependent. If domestic production is simply set free to clear markets, this answers both the issue of domestic pricing and, at the same time, eliminates any issue of special allocation. Similarly, it affects the macroeconomic issue, presumably by maximizing inflation and minimizing output and employment, while improving the balance of trade by reducing the demand for foreign oil. Many other interrelations could be described, but it will be best to discuss the issues separately, both because the exposition is simpler and because, in fact, the policy decisions were often made separately. While the macroeconomic impact is the main concern of this chapter, the other issues must be discussed first, since the results of that discussion are necessary to any analysis of the macro problem. An outline of the import problem that actually developed will serve as a background for the discussion of domestic policy alternatives.

Import Prices and the Embargo

While the Arab embargo on oil shipments to the United States was announced in mid October, its effect on actual shipments received came about gradually. The quantity of crude-oil imports was scarcely affected before December and did not reach its low point until February. Imports of crude oil started recovering noticeably in April, after the end of the embargo, and were at record levels by the third quarter of 1974. These developments during the period immediately preceding and following the embargo are summarized in Table 2-1.

Compared with shipments to the United States in the last quarter before the crisis, the third quarter of 1973, imports of crude oil dropped by 0.42 billion barrels (annual rate), or one-third, by the first quarter of 1974. The shortfall was even greater, perhaps 0.6 billion barrels (annual rate), when measured against an estimate of the trend of demand for imports before the embargo and price increases.[1] As shown in Column F in the table, the embargo came at a time of sharply rising dependence on foreign crude oil. Crude imports just before the crisis were equal to 33 percent of domestic crude production, up from 20 percent a year earlier.

Table 2-1 also shows how the pattern of imports of petroleum products differs from that of imports of crude oil. Although a strong seasonal pattern makes comparisons difficult, it appears that product imports were not much interrupted by the embargo, although they did rise dramatically in price. In the first quarter, they were 13 percent below what they had been a year earlier, 6 percent above what they had been two years earlier, and 6 percent below the level of the fourth quarter of 1974. While shipments of crude oil come directly from producing nations and the embargo on them could be enforced quite successfully, product imports come largely from refineries in Europe and the Caribbean. It was far more difficult for producing nations to enforce an embargo on goods reaching the United States indirectly through other countries.

Our knowledge of petroleum import prices, for both products and crude, is not completely reliable and cannot be analyzed as closely as quantities. This must be borne in mind in connection with Table 2-2, in which prices of imported crude oil provided from two sources—data from U.S. customs and the Federal Energy Administration (FEA)—and prices of major imported products, for the period immediately before the embargo to the present, are set forth. According to FEA data, average prices of imported crude oil rose noticeably every month from November through March, then leveled off during the remainder of 1974 at a delivered price around three times their pre-embargo levels. Prices in the customs data lag noticeably behind these FEA estimates until the second

1. For this purpose, the trend in import demand is calculated by subtracting the actual output of domestic crude oil and natural gas liquids from a 5 percent annual increase in total domestic petroleum product demand. Obviously, for longer periods, price changes and the price elasticity of domestic supply as well as demand would have to be accounted for.

Table 2-1. *A Comparison of U.S. Imports and Domestic Production of Crude Oil and Petroleum Products, 1972–74*

Billions of barrels, annual rate

Quarter or month	Imports			Domestic production		Ratios		
	Crude oil	Petro-leum products	Total	Crude oil[a]	Total product demand[b]	A/D	C/D	C/E
	A	B	C	D	E	F	G	H
Quarter								
1972:1	0.75	1.00	1.75	3.91	6.26	.193	.448	.280
2	0.70	0.84	1.54	4.01	5.59	.190	.384	.275
3	0.81	0.84	1.65	3.99	5.63	.203	.414	.293
4	0.91	1.01	1.92	3.95	6.42	.230	.486	.299
1973:1	1.07	1.22	2.29	3.88	6.68	.275	.590	.343
2	1.15	0.95	2.10	3.88	5.96	.297	.541	.352
3	1.29	1.03	2.32	3.84	6.13	.335	.604	.378
4	1.23	1.13	2.36	3.83	6.44	.320	.616	.366
1974:1	0.86	1.06	1.92	3.78	6.17	.228	.508	.311
2	1.33	0.90	2.23	3.73	5.84	.358	.598	.382
3	1.43	0.80	2.23	3.64	5.87	.393	.613	.380
4	1.45	1.04	2.49	3.58	6.07	.405	.696	.410
Month/1973								
July	1.28	0.98	2.26	3.86	5.98	.331	.585	.378
August	1.31	1.06	2.37	3.85	6.36	.341	.616	.373
September	1.27	1.06	2.33	3.81	6.07	.333	.612	.384
October	1.37	1.02	2.39	3.85	6.24	.355	.621	.383
November	1.26	1.24	2.50	3.84	6.73	.328	.651	.371
December	1.06	1.12	2.18	3.79	6.36	.278	.575	.343
Month/1974								
January	0.87	1.08	1.95	3.74	6.30	.233	.521	.310
February	0.82	1.08	1.90	3.84	6.34	.213	.495	.300
March	0.90	1.01	1.91	3.76	5.87	.239	.508	.325
April	1.19	0.99	2.18	3.76	5.81	.317	.580	.375
May	1.37	0.90	2.27	3.74	5.70	.366	.607	.398
June	1.44	0.81	2.25	3.68	6.01	.392	.611	.374
July	1.52	0.78	2.30	3.65	5.95	.416	.630	.387
August	1.41	0.83	2.24	3.66	5.90	.385	.612	.380
September	1.37	0.80	2.17	3.62	5.79	.378	.599	.375
October	1.44	0.93	2.37	3.63	6.00	.397	.653	.395
November	1.46	1.08	2.53	3.55	6.08	.411	.713	.416
December	1.45	1.11	2.56	3.56	6.11	.407	.719	.419

Source: Federal Energy Administration (FEA) data.

a. Including natural gas liquids. FEA data for production of crude oil exclude natural gas liquids, which amounted to 15 percent of total crude production in 1973. Output of crude oil plus natural gas liquids was estimated as 1.15 times production of crude oil in each period.

b. Total product demand, Column E, is estimated by the FEA by subtracting inventory charges from total refinery output plus product imports. For a variety of reasons, including the fact that domestic production of natural gas liquids in Column D was estimated by the author (see Footnote a), Column E does not equal Column C plus Column D less inventory changes.

Table 2-2. *Import Prices for Crude Oil and Major Petroleum Products*
Dollars a barrel

Quarter	FEA data Crude oil	Customs data[a]				
		Crude oil	Gasoline	Residual fuel oil	Distillate fuel oil	Jet fuel
1973:3	4.47	3.91	8.96	4.75	6.30	5.24
4	6.54	4.85	11.28	5.83	9.05	6.69
1974:1	11.59	9.60	17.79	10.78	14.53	11.11
2	12.93	12.48	20.63	12.43	15.09	13.37
3	12.65	12.55	16.09	12.34	14.79	14.11
		Increases from 1973:3 prices				
1973:4	2.07	0.94	2.32	1.08	2.75	1.45
1974:1	7.12	5.69	8.83	6.03	8.23	5.87
2	8.46	8.57	11.67	7.68	8.79	8.13
3	8.18	8.64	7.13	7.59	8.49	8.87

a. Customs data for crude oil and products are estimated by calculating the implied price from customs import data (value divided by quantity) and adding a constant markup to cover the costs of shipping, insurance, and distribution to bring the price to a level comparable to U.S. domestic prices. The amount of the constant markup was calculated as the difference between the U.S. domestic price and the implied customs price for the earliest period for which the two could be compared—in most instances the second quarter of 1973. With respect to crude oil, this adjustment added 95 cents to the implied customs price.

quarter of 1974. Prices of imported products, except for gasoline, followed much the same pattern as the FEA estimates of crude oil prices did.

If prices of petroleum products and crude oil are compared by means of the customs data,[2] product prices rose noticeably more than crude prices in the first two quarters after the embargo. These increases are compared directly in the bottom half of the table. One explanation for this behavior is that the prices of imported products responded more directly to the balance between supply and demand for products in the U.S. market than did crude prices. Except for residual fuel oil, these product imports represented only a small fraction of total U.S. demand for petroleum products and appear to have provided a marginal part of supply at a price well above the average product price during the period of shortage. An extraordinary price differential is especially apparent in gasoline imports, and gasoline was the product in clear excess demand during the first half of 1974. Imports of residual fuel, by contrast, represent about half of total U.S. demand for residual fuel. Longer-term contracts may help explain

2. The price levels in these data are adjusted as described in the footnote to Table 2-2. This adjustment adds a different constant amount to the prices shown in each column, so the changes within each column are the same as the changes implied in the original customs data.

the relatively modest rise in their price. Journalistic accounts of individual sales of residual shipments, however, particularly during the winter months, indicated that some extraordinarily high prices were being paid, and these may not be reflected in the prices derived from customs data.[3]

Pricing

Against this background of rising world oil prices and reduced shipments to the United States, a price policy was the most controversial decision that had to be made concerning domestic production. Three basic possibilities existed: a free-market solution, comprehensive price controls, and a partial control of prices. Each offered advantages and drawbacks.

FREE MARKETS. If the price had been left to be determined by market forces, the cartel-enforced rise in world prices would have driven up the price of domestic crude oil correspondingly. As a first approximation, this would have been the price at which imports in fact sold in the U.S. market. Under this condition, the free-market solution would have added a cost of roughly $8 a barrel on 3.8 billion barrels (annual rate) of domestic production (including natural gas liquids), or $30 billion, to the nation's petroleum costs, over and above the additional costs of imports. Although the demand for crude oil depends on the demand for petroleum products, given a fixed world price for crude oil, any adjustment to higher prices would take place entirely in the form of reduced consumption of petroleum products. Only if domestic production exceeded the quantity of products demanded would the price as well as the quantity adjust to the new market conditions.[4] The principal advantages of the free-market solution are the economist's conventional ones: economic efficiency in allocating petroleum products, the lack of any need for bureaucratic rules or

3. One major weakness of the customs data is that they are collected in a way that apparently offers no great inducement to accurate reporting of shipment values (or penalty for faulty reporting). Also, the values are reported on the basis of value at point of shipment (c.i.f.), and if an extraordinary price were obtained on delivery because of shortages during the embargo, it would not show up in the data. Of course, the same problem exists for the data on other imports, including crude oil. The major importers of crude oil, however, may have acted with somewhat more restraint for a variety of reasons.

4. If the price of delivered crude oil itself were variable, either because of the ability of importers to vary their markups over the cost to them from producer nations or because a sufficient reduction in import demand would influence the world oil price, the analysis would have to be amended.

monitoring, and the automatic incentive for expanding supply that high market-clearing prices would provide. The disadvantages are the huge cost increases to users of petroleum and the corresponding huge profits to producers of oil and collectors of royalty, which represent a politically unacceptable income windfall. Despite the fact that some of these implications could be altered by means of various tax schemes, the free-market solution was not accepted.

COMPREHENSIVE PRICE CONTROLS. At the other extreme from the free-market solution was the alternative of holding the price of domestic oil at its pre-embargo level or at some other level well below world prices. Legislation to control the price of oil did in fact exist during 1973, even before the oil crisis, so the issue was whether to decontrol rather than whether to impose controls. The embargo and rising world oil prices came at a time of gradually declining domestic production and sharply expanding reliance on imported oil. Freezing domestic prices, therefore, either by controlling the prices of petroleum products directly or by controlling crude prices and regulating the margin between crude and product prices, posed the serious risk that domestic production from existing reserves would be reduced as oil producers waited for prices to be allowed to rise.[5] The development of new reserves might also be reduced by uncertainty about future prices or other conditions of production. With total supplies curtailed by an embargo, moreover, there is no doubt that a price freeze would create excess demands for products, thereby making rationing necessary.

PARTIAL CONTROLS. The system actually developed involved a partial control of prices and was designed to maximize domestic production while compromising on the amount of the average price increase. A brief outline of the system that was adopted will suffice for the present analysis.

Four categories of crude oil were established: oil from stripper wells (wells from leases producing an average of less than ten barrels a day per well), "old" oil, "new" oil, and "released" oil. New oil was defined as all oil produced from a given lease in excess of what that lease had produced in the corresponding month of 1972. For every barrel of new oil, another barrel from the same lease was designated as released oil. Oil from stripper wells, new oil, and released oil were not subject to price control. The rest of production was designated old oil and its price was frozen, first

5. It goes without saying that exports could not be allowed if domestic prices were held well below world prices.

Table 2-3. *Crude Oil Prices by Category*

Dollars a barrel

Month and year	Imported		Domestic[a]			Average of domestic and imported
	Customs data	FEA data	Uncontrolled price[b]	Old oil[b]	Average	
1973						
July	3.80	4.35	3.79	3.79	3.94[c]	3.94[c]
August	3.91	4.47	3.86	3.86	4.01[c]	4.03
September	4.02	4.54	5.12	4.18	4.43[c]	4.35
October	4.18	4.91	5.62	4.11	4.47	4.47
November	4.45	6.49	8.50	4.17	5.43	5.44
December	5.92	8.22	9.51	5.25	6.49	6.54
1974						
January	7.36	9.59	9.82	5.25	6.72	7.46
February	9.73	12.45	9.87	5.25	7.08	8.57
March	11.72	12.73	9.88	5.25	7.05	8.68
April	12.55	12.72	9.88	5.25	7.21	9.13
May	12.46	13.02	9.88	5.25	7.26	9.44
June	12.45	13.06	9.95	5.25	7.20	9.45
July	12.74	12.75	9.95	5.25	7.19	9.30
August	12.60	12.68	9.98	5.25	7.20	9.17
September	12.32	12.53	10.10	5.25	7.18	9.13
October	12.24	12.44	10.74	5.25	7.26	9.22
November	12.18	12.53	10.90	5.25	7.46	9.41
December	12.25	12.82	11.08	5.25	7.39	9.28

a. Domestic prices are from the FEA, as is the average of domestic and import.

b. These are wellhead prices and average about 15 cents a barrel less than acquisition cost to refiners. All other prices shown, including the average domestic price, are based on acquisition cost to refiners.

c. Constructed by the author on the basis of FEA data.

at a wellhead price of $4.25 a barrel and then, after the middle of December 1973, at $5.25 a barrel.[6] According to estimates from the FEA, which are based on reports from producers, old oil as a percentage of total production rose from 60 to 67 percent during the first nine months of 1974.

The result of the dual-price system for domestic production resulted in three prices for crude oil delivered to U.S. refineries: the price for old oil, the price for domestically produced oil exempt from price control (new, released, and stripper output), and the price for imported oil. The available data, summarized in Table 2-3, indicate that imports and exempt domestic oil have been selling at substantially different prices. If these

6. These and all other prices cited in this paper are averages. Quality and locational differences exist for all petroleum products and for crude oil.

data are to be believed, they indicate that the markets for the different sources were isolated and that competition between imports of crude oil and domestic production was highly imperfect. Alternatively, it may be that the data on import prices are suspect and for the most part represent bookkeeping prices charged by integrated companies rather than the prices paid in "arm's length" transactions. Even if the latter alternative is correct, however, they still represent a cost of crude oil used for purposes of pricing products refined domestically.

The basic pricing rule connecting crude oil and petroleum products was a dollar-for-dollar pass-through of costs. Refiners were to calculate the average price per barrel of their crude oil input and the increase of this average over the price that they paid in mid May 1973, which was taken as the base price for this purpose. They could then raise product prices by this same amount per barrel. With some exceptions designed to encourage the production of certain products as part of the allocation effort, the basic rule was applied to each product separately, which meant that product price increases, in dollars per barrel, did not depend on the values of particular products. Other cost increases were allowed for, but they were insignificant in relation to the rise in crude oil prices. Two rules governing the timing of increases in the prices of petroleum products modified the basic pricing picture just described. Refiners could raise product prices with a lag (of no less than a month) after a rise in their crude oil costs.[7] During this period of rapidly rising oil prices, it was possible for refiners to store up increases in the cost of crude oil and use them to justify subsequent increases in the prices of petroleum products.

For most products, cost pass-through, with a short required lag for some products, was also the basic rule at subsequent levels of the distribution system. Gasoline retail prices were a notable exception to the cost-pass-through rule. Here a succession of special increases in dealer margins were permitted, which by spring amounted to 3 cents a gallon ($1.26 a barrel), or a total of about $3 billion at the normal volume of annual sales. This extra margin was designed to make up for the sharp decline in the volume of sales that occurred during the winter. It began to decline gradually at the end of April, but by September the margin was still nearly 2 cents a gallon above the level of the summer of 1973.

Again, details of these pricing rules are beyond the scope and purpose

---

7. Requiring a lag was presumably meant to take account of inventories that had been bought at lower prices.

of this chapter. The main point is that the total rise in the value of domestically refined products—which largely determined the macro effect of the oil crisis—was supposed to be a dollar-for-dollar pass-through of the rise in the value of crude oil from domestic and foreign sources with a short lag, plus a noticeable increment to gasoline prices at retail. The rise in the value of imported products was the other major contributor to the rise in the value of domestic sales. The wholesale value of total domestic product sales plus suppliers' inventory accumulation for the quarters immediately before and after the oil crisis is summarized in Table 2-4. The total is also broken down into these several main parts: the values of crude and product imports, domestic crude production, and refinery margins.

In Table 2-4, the total value of products from U.S. refineries is estimated by expanding the values of four principal products for which price and output data were available from the FEA—gasoline, middle distillates, residual fuel, and jet fuel. These products account for about 80 percent of the volume of total petroleum products. There is a negligible difference between the quantity of a refinery's output and input, so the projection of total products was made by adjusting the output of the four principal products up to the total input of crude oil to refineries and multiplying by 1.15 to allow for the contribution of natural gas liquids to total products. While this procedure should give a quite accurate estimate of the total volume of products each month, it may have a distinct bias as an estimate of the total value of products, since it assumes that the average value of the unobservable 20 percent equals that of the principal products. However, since the analysis rests entirely on changes in these values, a bias in estimating the level of the unobservable 20 percent should not be significant, particularly if the FEA rules of dollar-for-dollar pass-through were being observed for all products.

Sources of Rising Fuel Bills

While the value of products rose sharply in each quarter of the past year, the relation between the values of petroleum products and crude oil, quarter by quarter, is uneven. This makes it difficult to judge how well the pass-through rules were followed. Because of the required lag in passing crude-oil price increases into product prices, value added by refiners declined during the initial period of sharply rising crude-oil prices, as shown in Table 2-4. By the third quarter of 1974, it had recovered and

Table 2-4. *Rising Values of Petroleum Products and Their Sources*
Billions of dollars

Item	1973:2	1973:3	1973:4	1974:1	1974:2	1974:3
All products	39.70	43.16	50.87	63.11	73.70	77.02
Total imports	9.16	10.77	15.12	22.49	27.79	27.31
Crude oil	4.72	5.72	7.85	10.00	17.28	18.14
Petroleum products	4.44	5.05	7.27	11.49	10.51	9.17
Total domestic production	30.54	32.39	35.75	41.62	45.91	49.72
Crude oil	15.02	15.15	20.92	26.70	26.23	25.46
Value added by refiners	15.52	16.58	14.83	18.92	19.68	24.66

Source: Calculated by the author as described in the text.

had reached the highest level in any quarter shown. Since refiners could store up cost increases, the apparent rise in margins by the third quarter does not show that the pass-through rules were violated; but if the high level of margins were to be maintained indefinitely, it would indicate that they were.

Rising import values accounted for $16.5 billion of the increase in the U.S. value of petroleum products that occurred between the third quarter of 1973 and the third quarter of 1974—roughly half the total increase. Because it is difficult to reconcile all the import price increases on which these values are based (Table 2-1) with the increases indicated by producing nations, some part of these rising import values may reflect higher margins or shipping profits by major oil companies. The great bulk of the $16.5 billion, however—if not all of it—represents higher payments to foreign producing nations.

About $9 billion more of the rise in product values during this four-quarter interval reflects higher revenues of domestic producers of crude oil. The rise occurred despite a 5 percent decline in production. Since any increase in production costs over the period was negligible in comparison to the indicated rise in gross receipts, the $9 billion increase in revenues of crude-oil producers must have been divided between royalties and gross profits in the industry.

TAX CHANGES. In view of the huge increases in profits arising from higher oil prices, several proposals were made to change the tax status of the oil industry. The two seriously proposed within the government were the introduction of a windfall-profits tax and the removal of the percentage depletion allowance and other special tax incentives enjoyed by the oil industry. The windfall-profits tax was designed to tax away part of the difference between a relatively low base price and the market price; the higher the market price, the greater the fraction taxed away. The base price was to rise in time so that the tax would phase itself out automatically. The original proposal had the inherent weakness of encouraging producers to delay production in order to take advantage of the declining tax rate. During debate in Congress it acquired the additional weakness of granting a credit against the tax for reinvested profits. The tax has not been passed and its future is uncertain. The depletion allowance was brought to an end for all but independent operators (those producing less than 3,000 barrels a day) in legislation enacted in March 1975.

PRODUCTION INCENTIVES. Tax changes, such as the removal of the depletion allowance, would be expected to affect domestic supplies in the long run, though the size of their impact is hard to measure. Prices have risen so much that the past offers little indication of the quantity of supplies that will eventually be forthcoming. If most new fields were sufficiently profitable even in the absence of special incentives, the only limitation on potential supplies for many years might be the speed with which new fields can be brought into production. For the near future, however, speculation about prices or possible tax changes may be important in determining actual rates of production. The price freeze on old oil may be limiting current output from some fields, if producers are speculating that the freeze will be lifted as, from time to time, government spokesmen have suggested that it will be.

Allocation

During the months before the end of the embargo, the principal concern of energy policy was allocation of the essential supply of fuels and raw materials in such a way as to keep industry and essential services in operation and to keep homes from freezing. Achievement of these aims was sought in three ways: conservation was encouraged, and higher prices were allowed to reduce demand somewhat; use of gasoline by consumers was allowed to absorb the remaining shortfall in the total fuel supply; and rules for mandatory allocation were used to direct the available supply of crude oil and products. Broadly speaking, the goals were achieved: industry was not slowed for lack of fuels, homes were kept comfortable, and drivers suffered. How much each part of the effort contributed is harder to judge.

MANDATORY ALLOCATION. The President had authority to impose a program of mandatory allocation of crude oil and petroleum products under an amendment to the wage-and-price-control legislation passed in April 1973. He used it first to allocate propane, a fuel threatened by shortages arising from price regulations under Phase Three of the existing wage-and-price program. When the oil embargo went into effect, allocation programs were extended to cover crude oil and virtually all petroleum products. Although the precise rules varied from product to product, the main purpose in each case was to keep existing supply channels fairly evenly supplied.

Refineries with below-average supplies of crude oil were authorized to buy from refineries having supplies that were above average. Refiners were to supply each of their customers and distributors with petroleum products in proportion to the volumes supplied each in some base period —usually the previous year—and wholesalers were to treat retail customers in a similar way. For some products, certain users were recognized as having priority and could claim specified fractions of their total needs before the remaining supplies were apportioned among other users. Some supplies were also set aside for emergency allocation by state authorities to avert critical shortages and alleviate special hardships that might develop.

There is no way to evaluate compliance with the rules or to estimate the degree of inequity or inefficiency they caused. Producers, refiners, distributors, and customers, as well as politicians from states that appeared to have been short-changed by the rules, all complained. Whether the opposite approach—allowing the free market to allocate products—would have raised still more serious and legitimate complaints is impossible to know. Users worried a great deal about whether supplies would be adequate, whether their needs were to heat homes, run trucks and cars, or operate factories. But in fact, except with respect to gasoline for private consumption, widespread shortages did not develop.

THE GASOLINE PROGRAM. As part of its allocation effort, the Federal Energy Office (FEO, predecessor of the FEA) attempted to shift the composition of refinery output, particularly, away from gasoline. But despite shifts in refiners' margins designed to favor the production of distillates over production of gasoline, which started in December, and despite the existence of authority to dictate a gasoline production fraction to refiners, the statistics do not reveal that there was any shift. Table 2-5 shows the percentage of gasoline in total refinery output, both by month

Table 2-5. *Gasoline as a Percentage of Total Refinery Output*[a]

Month	Oil crisis (October 1973– September 1974)	Preceding year (October 1972– September 1973)	Two years earlier (January– September 1972)
October	45.2	47.9	...
November	44.8	47.3	...
December	43.7	46.2	...
January	44.7	45.3	47.1
February	45.9	43.9	46.0
March	46.2	43.9	45.4
April	46.9	45.5	45.4
May	45.4	47.6	45.7
June	46.6	47.3	45.9
July	46.0	47.7	48.5
August	46.2	47.4	48.2
September	46.1	45.9	47.5
Quarter	*1973:4–1974:3*	*1972:4–1973:3*	*1972:1–1972:3*
4	44.6	47.1	...
1	45.6	44.4	46.1
2	46.3	46.8	45.7
3	46.1	47.0	48.0

Source: FEA.
a. Refinery production of gasoline divided by crude oil plus natural gas liquids used by refiners.

and by quarter, for the period of the oil crisis and for the corresponding months and quarters one year and two years earlier. During the oil crisis, the fraction rose every month from December to March. It fell or remained the same in the corresponding months in earlier years.

Despite this failure to shift refinery output, a gasoline shortage was created. It seems to have arisen from two other causes: first, it came automatically, because almost no gasoline is imported directly, so the relatively uninterrupted flow of product, as opposed to crude, imports, during the embargo did not lessen the embargo's effect on gasoline supplies. Second, it came from the way in which the program of allocation and pricing of gasoline was operated.

The gasoline program essentially followed the general allocation rules just described. Suppliers allocated supplies to wholesale purchasers in the standard way, after setting aside a small amount to be allocated by state authorities and after supplying priority users with all or most of their current needs. In the case of gasoline, these special allocations were substantial, providing 100 percent of current requirements for agricultural production, emergency services, energy production, sanitation, telecommunications, and public transportation services, and an amount equal to 100 percent of 1972 purchases to other businesses. With total supplies limited, the shortfall to other consumers was substantial. Furthermore, total supplies were allocated by states, and, apparently because of the lopsided distribution of priority users, the shortfall fell quite unevenly. There were never any lines for gasoline in Minneapolis, for example.

Pricing policy encouraged some hoarding by retailers. Their prices could only be raised at intervals, generally at the beginning of a month. Anticipating an increase on March 1, stations closed down in late February, contributing to the worst period of shortage. Finally, the FEO held back on the total amount of gasoline made available for sale, which is perhaps understandable in view of the uncertain termination date of the embargo. Stocks of gasoline rose steadily from November through May. In January, the rise in stocks was about equal to 4.5 percent of the month's total consumption, and in February, it was equal to 1 percent of consumption. All in all, the handling of gasoline supplies was the worst part of the allocation program.

Conservation Measures

Both compulsory and voluntary conservation measures were a part of the response to the oil crisis. Steps taken included the following:

• Industries and utilities that were already using coal were prohibited from switching to oil, and the feasibility of substituting coal was considered by the FEO in allocating supplies of middle-distillate and residual fuels.

• The Interstate Commerce Commission adjusted routes in December for its regulated trucks and mail cars in order to conserve middle-distillate fuel.

• On 6 December the Federal Power Commission directed all utilities to aim at conserving 10 percent of their energy and to file a report on their conservation programs. Uses of coal, nuclear fuel, and hydroelectric capacity were to be maximized through existing power-transferring facilities.

• On 20 November the Civil Aeronautics Board authorized competing airlines to co-ordinate services, under Board supervision, in order to save jet fuel.

• A mandatory conservation program for government agencies was instituted, under which lighting, temperatures, and speeds of government vehicles were to be reduced, and the mileage driven in government vehicles was to be cut 20 percent from the mileage of the first quarter of 1973.

• Year-round daylight-saving time was legislated in late December, to take effect 6 January 1974.

• A nationwide speed limit of fifty-five miles an hour was legislated on 21 December; earlier, governors had been encouraged to impose speed limits of fifty miles an hour.

• On 21 November, the President asked gas stations to close voluntarily on Sundays in an effort to reduce pleasure driving. In January 1974, states were urged to institute their own gasoline-rationing plans and to include provisions that would prevent panic buying of gasoline by requiring that tanks be less than half full and that would stagger the hours of service stations. Many states adopted voluntary or mandatory rationing plans under which gasoline was to be sold for private cars only on odd- or even-numbered days, depending on license-plate numbers. On 17 March, before the futility of such schemes became apparent in the face of large excess demands, the end of the embargo was announced, and the gasoline shortage vanished.

How Did We Muddle Through?

Where does the credit go for the economy's surviving the winter of the oil embargo? A rough estimate made from Column E in Table 2-1 suggests that in the first quarter total consumption of all products fell about 12 to 15 percent below previous trends.[8] The general strategy of the energy planners succeeded to the extent that private use of gasoline was the hardest hit. And although there was no evidence of success in persuading refineries to cut the output of gasoline, it is possible that in an unregulated free market the refinery mix would have swung toward gasoline and that shortages of heating fuel would have developed. Conservation measures did have some effect, although their contribution probably amounted to no more than perhaps one or two percent of total demand. The slump in real GNP below trend accounts for some 3 to 4 percent. The price elasticity of demand must have provided a saving of a few per-

8. Total consumption was 12 percent below a growth rate of 5.0 percent a year earlier, and 15 percent below the growth rate of 6.7 percent that prevailed between the first quarters of 1972 and 1973.

cent more. The continuing flow of imports of petroleum products was crucial. And the good fortune of a warm winter helped.

Macroeconomic Effects of Oil Prices

The oil crisis has slowed the U.S. economy from the standpoint of demand rather than supply. Despite the obvious decline in business at gasoline stations during the winter of the oil embargo, industry and commerce have at no time been shut down for a lack of essential fuels. But total GNP has been increasingly depressed by the huge decline in aggregate demand that has resulted, both directly and indirectly, from higher oil prices.

The effects of the higher prices of petroleum products upon aggregate demand were manifested in the economy in a number of ways. The principal ones can be listed here.

Real-income effect. Real income was distributed away from consumers to oil-exporting nations and to the domestic petroleum industry. To the extent that the gainers in this redistribution have lower spending propensities than the losers, this shift in income depressed consumption and total demand.

Monetary policy effect. Since the demand for petroleum products is highly inelastic, especially in the short run, the increase in their price raised the value of expenditures on them, raised the nominal demand for money, and, in the absence of an accommodating increase in the money supply, raised interest rates.

Automobile demand effect. The demand for automobiles was depressed by the oil crisis and higher gasoline prices. Since car sales are partial substitutes for consumer saving, their decline added to the decline in total consumer spending that was predictable in the light of the real-income effect.[9]

Induced-inflation effect. To the extent that consumer prices influence wage increases, the petroleum price increase has resulted in higher wages and hence a new round of higher prices for other products. While the quantitative importance of this effect is less certain than that of the first

9. The Michigan quarterly econometric model, which explicitly contains an equation for automobile sales, treats them in this way. This treatment is also supported by other statistical research. See, for instance, Arthur M. Okun, "The Personal Tax Surcharge and Consumer Demand," *Brookings Papers on Economic Activity, 1:1971,* **pp.** 167–211.

three, it has further reduced the real value of the money supply and raised interest rates in some measure.

As each of these impacts was being felt in the economy, they interacted and induced further changes through familiar multiplier, accelerator, and interest-rate effects. In order to estimate the ultimate effects on the economy of the oil crisis plus induced effects, two large econometric models are employed—the Michigan quarterly econometric model and the quarterly econometric model of the Federal Reserve Board (FRB).[10] Each of these provides estimates of the total net effect on the economy that have come and are yet to come from the oil crisis. The monetary-policy and induced-inflation impacts are inherent in the structure of these models, once the initial price-level effects are fed into them. The impacts from redistribution of real income and from reduced automobile sales require some prior quantitative judgments from outside the models, as does the calculation of the impact of the oil price increase itself. I will describe the calculations I have made and will then discuss the results obtained from the models.

Real Income and Consumption

The impact of petroleum prices on the consumption deflator is allowed to determine the impact on total consumption in the models. This amounts to assuming that real consumption is determined by real disposable income and that changes in relative prices among different categories of consumption have no effect on total consumption (with the exception, as already noted, of automobile demand). This procedure does not directly estimate the separate effects of higher prices on consumption of petroleum products and on consumption of other commodities; however, the real decline in consumption of petroleum products that can be identified does reduce the weight of those products, and hence of their price increase, in changing the deflator.[11]

10. I am grateful to Jared Enzler of the research staff of the Board of Governors of the Federal Reserve System, for performing required projections on the FRB model and to Saul Hymans of the University of Michigan for doing the same on the Michigan model.
11. Imagine a price rise on a product with sufficiently elastic demand that its consumption became negligible. It would not then enter the deflator; real income, as measured, would be unaffected; and the welfare loss associated with that price increase would go undetected. If total real consumption were determined as described here, however, it would not change, and the multiplier and employment effects would be correctly calculated.

The impact of petroleum prices on the consumption deflator includes the weight of the prices of products bought directly by consumers—principally gasoline and heating oil—and the weight of the petroleum prices embodied in other goods and services purchased by consumers. While the price rise in all petroleum products, except possibly those bought by government, would be expected to pass through to consumer prices eventually, the pass-through would be much slower in some cases than in others. Rental car charges would be expected to pass rising gasoline costs through immediately, while at the other extreme, higher prices for fuel used in the production of machinery would be expected to appear in consumer prices only after a substantial delay. Allowance was made for such differences in the calculation of the consumption-deflator impact.

While these distributions of price increases were necessarily guessed at, they are not crucial to the results obtained from the macro model. What is most important is that all the increase in market value was redistributed to ultimate purchasers, whether it went initially into a direct-purchase item such as gasoline or indirectly into petrochemicals or fuel for factories, and that the portion assigned to consumers and government over a period of time is fairly accurate, since it is primarily on these two inputs that the macro models operate.

The result of the allocation of oil price increases is given in the table below. It shows what percentage of the consumption deflator is attributable directly to oil price increases in each of the five quarters since the oil embargo.

Percentage increment to consumption deflator
attributable to higher oil prices

1973:4	1974:1	1974:2	1974:3	1974:4
0.62	2.37	3.40	3.47	3.55

For all the quarters after 1974, the last quarter's increment, 3.55 percent, is used in the analysis of economic impact presented below. As of the fourth quarter of 1974, $7.8 billion of price increases attributable to oil had not been allocated to consumer goods. Of this, $1.8 billion was estimated to be in government purchases, and the remainder was in the price of investment goods or represented a cost increase that was still temporarily absorbed in reduced profit margins of industries using petroleum products.

These deflator impacts do not simply represent the rise in product values shown in Table 2-4 adjusted for inventory charges and retail margins. For each quarter they are calculated as the quantities used in that quarter multiplied by the difference between the prices of petroleum products in that quarter and prices in the third quarter of 1973. Calculated in this way, higher oil prices accounted for $38.6 billion of the value of nominal GNP less imports in the fourth quarter of 1974, with $30.8 billion of this total allocated to consumer purchases in that quarter. It is this increase in consumer prices, reflected in the deflator impacts shown in the table above, that operates on the macro models by reducing real disposable incomes and real consumption spending. In addition, the total rise of $38.6 billion in value affects the macro models by raising the transactions demand for money.

The increase of 3.5 percent in the consumption deflator and the consequent fall in real income are treated by the model just as if they had arisen from a general price increase. Of course, the price increase is in fact concentrated in petroleum products. To the extent that the sharply higher relative price of oil leads to reduced consumption of imported oil, part of the overall reduction in consumer purchases is absorbed by foreign producers of oil and is not reflected in reduced production of domestic goods and services. Unless some adjustments were made to the aggregate models, this development would be missed and the depressing effects of the oil price increase would be overstated somewhat. But, as is explained below in the discussion of net exports, a separate estimate was made of the reduction in oil imports for which their higher relative price was responsible, and this estimate was used as a basic input to the models.

Automobile Sales

Neither of the models uses gasoline prices—not to mention the uncertain availability of gasoline—in its explanation of automobile sales. I estimated the additional impact on total consumption from this source using the residuals (prediction errors using actual data) from the automobile demand equation in the Michigan model. Actual values of income, GNP, and other relevant variables are used in calculating these residuals, which were nearly zero in the two quarters just before the oil crisis. Thus it seems reasonable to identify later shortfalls of automobile sales below the levels predicted by the equation with the effects of the oil crisis and higher gasoline prices. The shortfall in automobile sales, at annual rates in 1958 prices, was $4.3 billion, $6.9 billion, $4.7 billion, and $3.7 billion in each of the four quarters, beginning with the fourth quarter of 1973. I assumed that half of these shortfalls in automobile sales was reflected in other consumer spending, while half was saved. This assumption probably represents an underestimation of the depressing effect on the economy in the early quarters, when most of the shortfall in automobile sales should be assigned to savings; but for the longer projection, it is in accord with historical evidence. Allowing for the special effects of previously announced increases in the prices of 1975 models on sales in the third quarter of 1974, the increase in savings attributable to higher oil prices was estimated as $3.5 billion (in 1958 prices) in the fourth quarter of 1974 and $2.7 billion (in 1958 prices) in each quarter thereafter.

Investment

A substantial part of the transfer of real income away from consumers goes into business incomes in the domestic petroleum industry and automatically appears in the models. The small portion of this that can be expected to flow back to personal incomes in the form of increased dividends is automatically allowed for in the models, as is the portion that goes to the government in higher taxes. The models also estimate what might be considered the normal investment response that would be generated by all the effects discussed above, including the shift in income to the petroleum industry. But they could not be expected to capture accurately any possible extraordinary impacts of the oil crisis upon investment spending.

In order to investigate the possibility that extraordinary impacts were important, I compared investment plans for 1974 existing in the fall of 1973 (survey responses received in November and early December) with plans for 1974 existing in the summer of 1974 (survey responses received in July and early August) as reported by the Bureau of Economic Analysis of the U.S. Department of Commerce. The four industries that seemed candidates for extraordinary investment responses were petroleum, other transportation (which includes oil and gas pipelines), mining, and electric utilities (where downward revisions were expected). The extraordinary investment effects detectable in this way balanced out almost to zero, as shown in Table 2-6.

Because total spending by utilities is so large, the retrenchment in the

Table 2-6. *Investment Spending Plans for 1974*

Billions of dollars

| Industry | Date of plans | | |
	Late 1973	Mid 1974	Change
Petroleum	6.89	7.58	+0.69
Other transportation	1.62	2.23	+0.61
Mining	3.14	3.07	−0.07
Electric utilities	18.81	17.85	−0.96
Total	30.46	30.73	+0.27
Total with allowance for inflation[a]	30.46	30.42	−0.04

Source: U.S. Department of Commerce, *Survey of Current Business*, Vol. 54 (January and September 1974).

a. Plans of mid 1974 were reduced by 1 percent to allow for higher prices in the plans reported than from those reported nine months earlier.

plans of utilities already under way fully offsets expanded plans in other sectors. Even in current dollars, the total rise shown in the table is very small, and allowing only 1 percent for higher prices in the later reports makes the total change negative. Therefore, no extraordinary real investment impact was introduced into the models.

Net Exports

The other substantial transfer of income from consumers goes to foreign oil-producing nations. Observers all agree that given enough time exports to oil producers will rise substantially; there is less agreement on how big the eventual rise will be, and there is widespread disagreement on how big a rise can be expected in the short run. For the present purpose, estimates of the net export impact were made after distinguishing three types of effects arising from higher oil prices: the effect of the oil price rise on the volume of U.S. oil imports; the effect on U.S. exports of trade surplus improvements by oil producers and trade surplus deteriorations by other oil consumers; and the effects on U.S. imports other than oil induced by declines in GNP resulting from the oil crisis.

The first two effects were calculated to fit in roughly with worldwide projections of oil trade made by the editors of this volume. In contrast to a rising trend in oil imports that could have been expected before prices rose sharply, the volume of oil imports to the United States was projected to level off for the next few years, thus falling behind the earlier trend by

Table 2-7. *Net Export Effects of Higher Oil Prices*[a]

Imports or exports	1974	1975	1976	1977
Oil imports (billions of barrels)	−0.5	−0.7	−1.2	−1.5
Oil imports (billions of dollars)	+14.0	+13.9	+11.3	+9.5
Net exports (billions of dollars)	−12.0	−11.3	−6.9	−3.4
Net exports (billions of dollars, 1958 prices)	+3.0	+4.3	+7.0	+9.2

Source: Estimated by the author.

a. The table shows the difference in each year attributable to the higher level of world oil prices before induced effects from various levels of U.S. economic activity are allowed for.

increasing amounts during the 1974–77 period. By 1977, annual imports are projected to be 1.5 billion barrels below the low-price trend. At a world oil price of $10 a barrel, this still leaves the United States paying $9.5 billion more for oil imports in 1977 than it would have paid without the oil price increase.[12]

In response to trade surplus changes arising from oil, U.S. exports are projected to rise gradually, in comparison with an extrapolation of their former trend, to a level $6 billion higher than they would otherwise have been in 1977. This total gain in exports is the balance between increased exports to oil-exporting nations, offset to some extent by declining exports to developing nations that must conserve foreign exchange to pay for high-priced oil.

A complete projection of these two effects is given in Table 2-7. On the assumption that, after 1974, the prices of oil and U.S. exports move proportionately and by only small amounts, the increment to net exports projected there is also approximately the projected increment in current prices each year.

The third effect on U.S. net exports—the reduction in U.S. imports of other goods induced by changes in real GNP—is left to be estimated by

12. The estimates of the reduction in the volume of oil imports, compared to base projections, were revised subsequent to the preparation of this chapter and are presented in Table 1-1. The revisions lowered the projected reduction in imports and therefore raised the estimate of the additional import bill arising from higher oil prices (see Table 1-2). For the reasons offered in Chapter 1, the revised estimates are on the pessimistic side. In any event, incorporating them into this chapter would not have changed the macroeconomic results substantially. Higher oil imports are a substitute for domestic production of high-priced new oil or for other domestic sources of energy. Given the investment assumptions made by Perry, the macroeconomic results are not highly sensitive to the question whether higher economic rents are paid by consumers to domestic or to foreign producers.—EDS.

means of the macro models. The response of net exports to relative prices in these models was suppressed, since the effect of the oil price increase was projected independently and used as an exogenous impact in the models, as discussed above. Allowing the net exports in the models to respond only to changes in U.S. real GNP amounts to disregarding the real GNP effects induced by oil prices in other countries as these affect U.S. exports. This is the correct procedure if one assumes that other countries are free to stabilize their economies to whatever degree they choose. In the discussion of the projections, I consider the alternative assumption that real GNP in other industrialized countries is affected by the oil price increase in much the same way as it is in the United States.

Model Projections

The restrictive effects of the oil crisis on the economy have been large and are still growing. The econometric models into which the data described in the previous sections were fed both confirm this. In Tables 2-8 and 2-9 the results for the FRB model and the Michigan model respectively are summarized for the quarters since the oil crisis and are extended through the end of 1977. The tables show the difference that the rise in oil prices after the embargo has made in key macroeconomic variables such as GNP and its components, real output, unemployment, and average prices. In each model, these differences are calculated by subtracting a control solution of the model from its oil-crisis solution. The inputs for the two solutions differ by the oil-price impacts that have just been described. The models then provide estimates of the changes in major economic variables induced by these impacts.

Fiscal and monetary policy are kept the same in the no-crisis and control solutions of the models according to the following definitions: In both models, tax laws are held constant, and discretionary federal expenditures are kept the same in current dollars. Changes in revenues and expenditures that are induced by the level of economic activity, such as unemployment payments, are determined by equations in the models, as are state and local revenues and expenditures. The money supply is kept the same for both solutions in the FRB model, while the level of unborrowed reserves is kept the same in the Michigan model. These two assumptions are not exactly equivalent, as fixing unborrowed reserves can be expected to lead

Table 2-8. *The Effects of the Oil Crisis: Federal Reserve Board Econometric*
Dollar amounts in billions

Economic variable	1973:4	1974:1	1974:2	1974:3	1974:4	1975:1
Real GNP and components (1973 dollars)						
GNP	−5.4	−16.0	−24.1	−29.6	−35.6	−37.0
Consumption	−5.8	−15.4	−19.9	−22.2	−26.5	−26.8
Gross private domestic investment	0.1	−2.7	−7.2	−9.4	−10.3	−12.1
Inventory change	0.4	−1.7	−5.4	−6.2	−5.6	−6.0
Business fixed investment	−0.2	−0.9	−1.7	−2.8	−4.1	−5.3
Residential construction	0.0	−0.1	−0.0	−0.4	−0.7	−0.8
Net export	0.8	4.0	5.8	5.4	4.4	5.6
Government purchases	−0.5	−2.0	−2.8	−3.4	−3.2	−3.8
Disposable personal income	−5.7	−21.5	−31.6	−33.1	−32.3	−30.1
Current-dollar GNP and components						
GNP	−3.6	−6.9	−11.1	−14.0	−21.4	−21.7
Consumption	−1.0	1.9	4.1	2.6	−0.8	−1.5
Gross private domestic investment	0.6	−0.4	−3.9	−5.9	−7.0	−9.9
Inventory change	0.3	−1.7	−5.2	−6.2	−5.5	−6.6
Business fixed investment	0.3	1.1	1.1	0.4	−1.0	−2.7
Residential construction	0.0	0.2	0.2	−0.1	−0.4	−0.6
Net export	−3.2	−7.9	−10.5	−10.1	−13.5	−10.1
Government purchases	−0.1	−0.5	−0.8	−0.6	−0.1	−0.2
Disposable personal income	−1.5	−5.1	−7.7	−8.0	−6.4	−5.0
Corporate profits and IVA	−1.7	−0.0	0.2	−1.6	−11.2	−12.9
Personal saving rate (percentage points)						
Other economic variables						
Unemployment rate (percentage points)	0.1	0.4	0.8	1.0	1.1	1.2
GNP deflator (percent)	0.1	1.0	1.1	1.4	1.5	1.7
Compensation per man-hour (percent)	−0.1	−0.2	−0.1	0.1	0.4	0.7
Federal government revenues	−0.4	1.4	−0.3	−1.7	−6.1	−5.2
Federal government expenditures	0.1	0.2	0.7	1.6	2.5	3.8
Federal government surplus	−0.5	1.1	−1.1	−3.3	−8.6	−9.0
Treasury bill rate (percentage points)	−0.1	0.1	0.1	0.1	−0.3	−0.0

a. The table shows differences in the economy attributable to the oil crisis assuming that fiscal policy and

Table 2-9. *The Effects of the Oil Crisis: Michigan Quarterly Model*[a]
Dollar amounts in billions

Economic variable	1973:4	1974:1	1974:2	1974:3	1974:4	1975:1
Real GNP and components (1973 dollars)						
GNP	−3.9	−9.1	−15.1	−21.5	−33.2	−39.8
Consumption	−4.6	−11.7	−16.4	−20.1	−26.7	−29.9
Gross private domestic investment	−0.1	−1.4	−3.2	−5.5	−7.9	−11.8
Inventory change	−0.1	−1.2	−2.2	−3.2	−4.2	−6.1
Business fixed investment	−0.0	−0.2	−0.9	−2.0	−3.2	−4.9
Residential construction	−0.0	−0.0	−0.1	−0.3	−0.5	−0.8
Net export	1.2	5.1	6.3	7.4	7.1	8.4
Government purchases	−0.4	−1.1	−1.8	−3.3	−5.7	−6.5
Disposable personal income	−6.8	−23.0	−30.5	−31.7	−39.4	−43.2
Current-dollar GNP and components						
GNP	−1.3	3.4	4.0	3.6	−0.6	1.0
Consumption	0.4	7.4	11.3	10.5	9.3	11.3
Gross private domestic investment	1.1	3.1	3.3	1.7	0.8	−1.9
Inventory change	0.0	−0.7	−1.6	−2.9	−3.7	−5.8
Business fixed investment	1.1	3.7	4.5	4.0	4.1	3.7
Residential construction	−0.0	0.1	0.4	0.6	0.4	0.2
Net export	−2.8	−7.1	−10.6	−8.6	−10.7	−8.4
Government purchases	0.0	0.0	0.0	0.0	0.0	0.0
Disposable personal income	−1.2	−1.8	−0.4	1.1	0.9	1.3
Corporate profits and IVA	0.3	5.5	3.4	0.6	−1.4	−1.8
Personal saving rate (percentage points)	−0.2	−0.9	−1.2	−1.0	−0.9	−1.0
Other economic variables						
Unemployment rate (percentage points)	0.1	0.2	0.3	0.5	0.8	1.0
GNP deflator (percent)	0.2	1.0	1.5	1.9	2.6	3.3
Compensation per man-hour (percent)	0.0	0.5	2.0	3.3	3.4	4.0
Federal government revenues	n.a.					
Federal government expenditures	n.a.					
Federal government surplus	−0.3	2.0	2.4	2.1	0.2	0.7
Treasury bill rate (percentage points)	0.0	−0.0	0.0	0.0	0.1	0.0

a. The table shows differences in the economy attributable to the oil crisis assuming that fiscal policy and
n.a. Not available.

Model[a]

1975:2	1975:3	1975:4	1976:1	1976:2	1976:3	1976:4	1977:1	1977:2	1977:3	1977:4
−38.0	−39.2	−40.0	−39.8	−39.5	−39.6	−39.9	−40.6	−41.8	−44.0	−47.0
−29.1	−31.5	−34.2	−36.6	−38.9	−40.7	−42.4	−43.6	−44.8	−46.7	−48.9
−11.4	−11.1	−10.0	−8.6	−7.0	−6.2	−5.7	−5.4	−5.1	−5.2	−6.7
−4.2	−3.4	−2.2	−1.2	−0.2	0.3	0.6	0.3	−0.1	−0.7	−1.9
−6.3	−6.8	−6.8	−6.4	−5.7	−4.9	−4.1	−3.4	−2.7	−2.2	−2.0
−0.9	−1.0	−1.0	−1.0	−1.2	−1.7	−2.2	−2.4	−2.3	−2.3	−2.7
6.7	8.0	9.5	11.3	12.9	14.4	15.6	16.5	16.6	16.9	17.1
−4.2	−4.6	−5.3	−5.9	−6.4	−7.0	−7.5	−8.0	−8.6	−9.0	−9.2
−32.9	−35.2	−38.8	−41.5	−44.3	−46.3	−48.2	−50.1	−51.9	−54.7	−56.1
−20.4	−18.5	−15.7	−11.7	−7.2	−3.0	1.0	4.5	7.2	7.3	5.1
−2.6	−3.0	−3.1	−2.8	−2.2	−1.4	−0.2	1.3	2.0	1.6	0.3
−9.4	−9.1	−7.5	−5.5	−3.3	−1.8	−0.5	0.6	1.9	2.6	1.3
−4.9	−4.0	−2.7	−1.5	−0.4	0.2	0.6	0.4	0.1	−0.5	−2.0
−3.9	−4.4	−4.2	−3.5	−2.4	−1.2	0.1	1.4	2.5	3.0	4.3
−0.6	−0.7	−0.6	−0.4	−0.5	−0.9	−1.2	−1.2	−0.7	−0.4	−0.9
−8.1	−6.2	−4.6	−2.7	−0.9	0.9	2.4	3.5	4.4	4.3	4.0
−0.2	−0.3	−0.5	−0.8	−0.7	−0.8	−0.8	−0.9	−1.1	−1.1	−1.1
−4.6	−4.9	−5.6	−5.1	−4.2	−3.1	−1.8	−1.1	−0.7	−1.6	−0.8
−13.2	−11.1	−7.3	−3.1	1.3	4.7	7.8	10.6	13.4	13.5	11.1
1.3	1.4	1.6	1.7	1.8	1.9	2.0	2.1	2.2	2.3	2.4
1.9	2.2	2.4	2.8	3.0	3.3	3.7	4.0	4.2	4.4	4.6
1.0	1.2	1.4	1.6	1.8	2.1	2.3	2.6	2.8	2.9	3.1
−4.8	−4.0	−2.7	−1.1	0.7	2.3	3.8	5.0	6.0	5.6	4.5
4.0	4.3	4.4	4.6	4.8	5.0	5.3	5.6	5.9	4.5	4.8
−8.8	−8.3	−7.0	−5.6	−4.1	−2.7	−1.5	−0.6	0.1	1.1	−0.3
0.2	0.4	0.8	1.2	1.5	1.8	2.1	2.3	2.6	2.7	2.6

the growth of the money supply would have been the same without it.

1975:2	1975:3	1975:4	1976:1	1976:2	1976:3	1976:4	1977:1	1977:2	1977:3	1977:4
−47.2	−54.4	−60.1	−64.6	−67.5	−68.7	−68.4	−67.3	−64.8	−61.4	−57.5
−34.0	−38.4	−41.8	−44.9	−47.5	−49.4	−50.4	−51.1	−51.1	−50.4	−49.5
−14.8	−17.8	−20.3	−22.0	−23.1	−23.2	−22.5	−21.2	−19.2	−17.0	−14.4
−7.0	−8.1	−8.8	−9.1	−9.4	−9.1	−8.6	−7.3	−6.7	−5.6	−4.4
−6.8	−8.6	−10.4	−11.9	−13.0	−13.8	−14.1	−14.0	−13.6	−12.8	−11.7
−1.0	−1.1	−1.1	−1.0	−0.7	−0.3	0.2	0.6	1.1	1.4	1.7
9.7	11.1	12.4	13.8	15.2	16.4	17.5	18.1	18.5	18.8	19.1
−8.1	−9.4	−10.4	−11.5	−12.1	−12.5	−13.0	−13.1	−13.0	−12.8	−12.7
−50.0	−54.5	−56.7	−60.3	−62.7	−63.6	−64.7	−65.6	−65.0	−64.0	−63.1
2.6	3.3	3.3	4.6	5.7	7.0	9.4	12.2	15.2	18.2	21.4
12.2	12.1	11.9	12.1	12.1	11.7	11.9	12.1	12.4	12.7	13.1
−3.3	−4.9	−6.5	−7.5	−8.5	−7.5	−6.1	−4.9	−2.3	0.2	2.9
−6.7	−7.6	−8.4	−8.7	−8.6	−8.5	−7.8	−6.9	−5.9	−4.7	−3.4
3.3	2.4	1.1	−0.0	−1.0	−1.4	−1.2	−0.8	−0.2	0.8	2.0
0.1	0.3	0.8	1.2	1.7	2.4	2.9	3.3	3.8	4.1	4.3
−6.3	−3.9	−1.6	0.0	1.5	2.8	3.6	4.5	5.1	5.3	5.4
0.0	0.0	0.0	0.0	0.0	0.0	0.0	0.0	0.0	0.0	0.0
2.0	1.5	2.5	2.8	3.0	3.7	3.8	4.1	5.2	6.0	6.4
−1.3	0.2	−0.6	−0.6	0.1	0.3	2.3	4.2	5.7	8.0	10.1
−1.0	−1.0	−0.9	−0.8	−0.8	−0.7	−0.7	−0.7	−0.6	−0.6	−0.6
1.3	1.6	1.7	1.8	1.9	1.9	1.8	1.7	1.6	1.5	1.3
3.9	4.4	4.8	5.1	5.3	5.4	5.5	5.5	5.4	5.3	5.2
4.3	4.2	4.8	5.1	5.2	5.5	5.4	5.3	5.3	5.2	5.1
1.3	1.6	1.9	2.2	2.7	3.2	4.3	5.4	6.8	8.5	10.1
0.1	0.2	0.3	0.4	0.6	0.7	0.8	0.9	1.0	1.0	1.0

the growth of the money supply would have been the same without it.

to a slightly larger money supply if GNP is greater, but they are equally plausible definitions of maintaining the same monetary policy.

Real Demand

For the first year and a half after the oil crisis, similar depressing effects are attributed to higher oil prices in both models. By the first quarter of 1975, real GNP (expressed here and in the discussion that follows in terms of 1973 prices) is $37 billion to $40 billion (about 3.2 percent) lower. Weakness in real consumption spending accounts for most of the shortfall of GNP in these early quarters and throughout the projections.

By the second half of 1975, the effect of oil prices on real activity becomes substantially larger in the Michigan model than in the FRB model. In the Michigan model, real business fixed-investment spending falls increasingly below the no-crisis projection, reaching a level $9 billion below it in the second half of 1975 and averaging $13 billion below in the 1977 quarters. In addition, inventory investment is depressed by $8.5 billion in the second half of 1975 and is still down by $6 billion in the year 1977. By comparison, inventory investment has started to stabilize by late 1975 in the FRB model and is virtually unaffected by the oil crisis thereafter. Real business fixed investment stabilizes quickly too. After being depressed by $6.8 billion in the second half of 1975, this sector is progressively less affected by oil in subsequent quarters in the FRB projection. In general, the Michigan model shows a stronger and longer-lasting accelerator pattern in its investment sectors, which, in turn, contributes to further weakness in disposable income and consumption spending.

Both models show real government purchases falling behind as a result of higher oil prices and their effects on real activity.[13] In both models, the combination of more inflation and lower levels of real activity and income induce lower levels of real government purchases. The borrowing power of state and local governments is limited, and their expenditures

13. Inadvertently, the effect of higher oil prices on government purchases was made slightly different in the two models. In the Michigan model, purchases are assumed to be unchanged in nominal terms as a result of the oil prices. In the FRB model, purchases are assumed to rise about $1 billion by the end of 1975 to offset in part the effect of higher oil prices on government activity. The projections shown in Tables 2-8 and 2-9, however, are dominated by the effects on the government sector as estimated by means of the two models, and not by this small difference in the treatment of the impacts.

are constrained by tax revenues or aid from the federal government. In the federal sector, the weakness in real purchases reflects what may be unduly restrictive assumptions about the reactions of the budget to inflation. Discretionary expenditures are assumed to be the same in nominal terms in the oil-crisis and control projections (except for the minor accommodation to higher oil prices themselves in the FRB model). In the face of growing weakness in the economy, one might expect discretionary expenditures in the federal budget to be increased as a stabilization measure. The projections here must be viewed as what would happen if such stabilization policy was not undertaken.

Both models show a substantial stabilizing influence from real net exports. By the last quarter of the projection, these are higher by $17 billion to $19 billion, providing a considerable support to the economy in the face of lower demands from other sectors, and even by the end of 1975, real net exports are estimated $9.5 billion to $12.5 billion higher by the two models. Part of these increments represents the direct effects of higher oil prices in increasing imports by oil-producing countries from the United States, described earlier. But a part also comes because reduced real activity reduces import demand. In contrast to the case of government expenditures just discussed, a reasonable alternative assumption for the net export part of the model would have led to less support in a weaker economy. If the model had been run on the assumption that other nations experienced a downturn similar to that in the United States as a result of higher oil prices and if this were modeled as an endogenous part of the oil crisis rather than a failure of stabilization policies in these countries, then the stabilizing effect of lower import demand on the U.S. economy would be countered by a comparable destabilizing effect of lower exports as other nations went through a recession-induced reduction in their own imports. This alternative is discussed further below.

Wages and Prices

How much wages, and therefore other prices, would rise in response to higher oil prices is one of the important and still unsettled questions connected with the oil crisis. The two models give noticeably different answers to this question. In the Michigan model projections, compensation per man-hour responds quickly to the rise in consumer prices resulting from the oil price increase. This effect clearly dominates the restraining influence of weaker labor markets, and by the end of 1974, compensation per man-hour is 3.4 percent higher than in the no-crisis solution. These rising labor costs in turn induce higher product prices for goods and services produced domestically and the GNP deflator follows these higher labor costs upward with only a brief lag. By the end of the projection period, compensation per man-hour is 5.1 percent higher and the GNP deflator is 5.2 percent higher as a result of the oil crisis. Only 1.4 percentage points—about one-fourth—of this increment to the deflator is attributable to higher domestic oil prices themselves.

The FRB model projects a considerably slower response of wages to higher oil prices and less induced inflation from this source even by the end of the projection. In this respect the actual structure of the models differs even more than is indicated by the projections of wages and prices, since the FRB model projects less of a loss of real output and a lower level of unemployment. Yet while the wage-price spiral is dampened more in the FRB model, it is far from negligible. Rising slowly each quarter, compensation per man-hour is 3.1 percent higher by the end of the projection period, as a result of higher oil prices and the subsequent wage-price spiral set off by them.

These wage-price effects are not great when viewed in terms of inflation rates. In the FRB model, they amount to an acceleration of approximately 0.9 percentage point in both wages and prices in each year after the end of 1974. In the first year, the wage acceleration is negligible. In the Michigan model, the increment to wage rates is 3.4 percentage points the first year, but only about 0.6 percentage point a year thereafter. However, the higher wage and price levels do add to the depressing effects of higher oil prices themselves in a model in which it is assumed that policy—particularly monetary policy—remains unchanged. The real value of the money supply and of federal expenditures is reduced by this inflation, adding to downward pressure on the economy.

Alternative Projections

The results of the macro models came from particular assumptions about prices, policies, imports, and the like. It is worth considering some alternative assumptions and the difference that they could be expected to make. There are too many plausible alternatives to make it feasible to re-run the macro models for each one. Instead, the way in which they would alter qualitatively the central projections that were presented in Tables 2-8 and 2-9 and discussed above will be explored in this section.

OIL DEMAND. The level of future oil imports will depend on policies toward imports as well as on the response of domestic demand and on the supply from domestic sources. None of these can be estimated with any precision, so the projection of the amount by which oil imports are reduced by higher oil prices is quite uncertain. To the extent that the alternative to imports is new domestically produced oil priced at the same level as imported oil, it makes very little difference in the calculated impact on the economy. In both cases, income is transferred from consumers to producers, and in both cases, the incremental demand on the part of producers is small in relation to the transfer of income. Similarly, while the effect of higher prices on the total demand for oil cannot be projected accurately, the range of uncertainty about quantities consumed is small in relation to the huge price increase that has occurred. This follows from the low elasticity of demand for oil that is widely acknowledged, at least for the relatively short run, with which the present analysis is mainly concerned. As a result, estimates of the total depressing effect on the economy are not sensitive to modest errors in the projection of demand. Thus, the estimates of the impacts on the economy are not highly sensitive either to the total amount of oil demanded or to the fraction of that total that is supplied by imports rather than domestic production.

OIL PRICES. Quantitatively, greater importance attaches to the uncertainty in the projection of future prices. The macro models worked from the assumption that the deflator impacts that could be estimated for the fourth quarter of 1974 would be maintained in future quarters. Should the cartel price break, this assumption could prove to be too depressing, while if the price were increased—say to keep pace with other world prices, as the cartel members have suggested—then future price effects would be greater than those projected here, and the depressing effect would be correspondingly greater.

The price effect could well be greater than that estimated here even if world prices are simply maintained at present levels. Most of the policy options being considered for pricing domestically produced oil would narrow or close the gap between average domestic prices and world prices for oil. If price controls were simply lifted, the price of all domestically produced oil would rise to the world price; at late 1974 price differentials, this would reduce real consumer incomes by another 1.5 percent. Combined with higher world oil prices, or with tariffs on imports to the United States, such a policy could add as much to the cost of petroleum products in the future as was added during the first year following the oil crisis. Whether or not any such future increase were captured in higher taxes on the oil industry, it would be necessary to offset the further depressing effect on the economy through tax reductions and faster growth of the money supply if the same kinds of economic impacts projected in Tables 2-8 and 2-9 were not to be repeated and added to the effects shown here.

OTHER FUELS. The depressing effects of the oil crisis on the economy are underestimated in the projections to the extent that they make no allowance for price increases in other fuels that are competitive with oil. Prices of coal and natural gas have risen with oil prices. While on a BTU-equivalent basis, the importance of these two fuels is comparable to that of oil, the aggregate importance of increases in their prices has not been nearly so great. Much of natural gas production is price-controlled, and both coal and gas are generally sold under long-term contracts. As time passes, price increases for these two alternative fuels will become increasingly significant to the aggregate economy. For the future, the extent to which they rise will depend in part on policies still to be made, particularly in the case of natural gas.

RECESSIONS ABROAD. It was noted above that the macro-model projections did not allow for any oil-induced recession in other countries to show up in reduced exports from the United States, on the grounds that the size of such depressing effects on foreign economies depended on the stabilizing actions that they took. For the U.S. case, policy has been treated as unaffected by the oil prices. If the same assumption is made for other countries, then the depressing effects of oil abroad should be treated as acting on the United States through a reduced demand for U.S. exports. Making such an assumption would reduce some of the stabilizing effect that a growing net export surplus has on U.S. economic activity in the basic projections presented here.

Subtracting the direct effects of the oil price increase on net exports discussed earlier from the ultimate net export increment shown in Tables 2-8 and 2-9 yields the induced reduction in imports in the model projections. During the last half of the projection period in both models, the induced fall in imports amounts to about 8 percent of the fall in GNP. If, in contrast to what was done, we had assumed a similar fall in U.S. exports because of oil-induced recession abroad, the ultimate effect would add perhaps 20 percent to the indicated shortfall in GNP that has been projected here. This amount is hardly negligible. But neither is it so large that it changes the qualitative character of the projections shown in Tables 2-8 and 2-9.

Conclusion

Higher oil prices, coupled with the failure of fiscal and monetary policies to offset their depressing effect on total demand, have been the primary cause of the present recession. They have also added noticeably to inflation in recent quarters. The econometric model results presented earlier offer quantitative estimates of these effects. By the end of 1974, the depressing effect of higher oil prices had reduced real GNP (1973 prices) by about $35 billion and added about 1 percentage point to the unemployment rate. Higher oil prices had directly added 3.5 percent to the consumption price deflator and were projected eventually to add roughly that much again through wage-price spiral effects. The possible amendments to these results considered in the previous section indicate they may understate the effects that oil prices have had and will continue to have on the aggregate economy. In particular, three considerations that were not allowed for in the econometric model estimates point in this direction: policies in other industrial nations have not offset oil-induced recessions, and these recessions abroad are depressing the U.S. economy further; average coal and gas prices are rising steadily in response to oil prices, with corresponding depressing aggregate effects; and average oil prices themselves are likely to rise noticeably beyond their levels of late 1974, in part by the deliberate design of U.S. policies.

This analysis clearly warns of the need to offset the depressing effects of higher energy prices on the economy. It also points out that higher energy prices aggravate inflation both through their direct, one-time effect on the average price level and indirectly to the extent that this direct effect sets off faster wage increases. An incomes policy that did not allow wages to chase higher energy prices upward could eliminate this indirect inflationary effect. Tax reductions (or subsidies) that directly reduced prices—such as reductions in excise taxes or employers' payroll taxes—could offset both these inflationary effects and the loss of purchasing power as well. But conventional fiscal measures, as well as conventional monetary policy, can moderate inflation only through lowering total demand and raising unemployment and excess capacity. This is the course that fiscal and monetary policy followed throughout 1974 when it failed to respond to higher oil prices.

Judging from their earlier statements, it is doubtful that policy makers intended or expected a recession as deep as that experienced as a result of these policies. And so long as the economy continues to operate below target levels of employment and utilization, the need to undo some of the depressing effects of higher oil prices will continue to confront policy makers. Finally, should energy prices rise further in the future, the stabilization problem of 1974 will arise anew. If policy once again fails to respond—by restoring lost purchasing power and allowing a more rapid expansion of monetary aggregates so as to keep interest rates from rising—real economic activity will be pushed down once again, this time from levels that are clearly depressed already.

CHAPTER THREE

Western Europe

GIORGIO BASEVI
University of Bologna

IN ORDER TO ANALYZE the major problems faced by Western Europe after the oil crisis, one must first trace the effects that higher oil prices would have had on major macroeconomic variables during the period 1974–77 if economic policy had been neutral—if, that is to say, neither offsetting nor reinforcing policy reactions had been superimposed on the adjustment path of the economic system. On the basis of the results thus obtained, one can then analyze the responses of policy authorities to the crisis and recommend broad lines of policy for the future that should serve to carry the economy along a reasonable path of adjustment. Finally, one can take up the long-range problems—the identification of the economic consequences of the oil crisis for Western Europe after the reabsorption of the initial imbalance and the policies that these consequences will require.

IN PREPARING this study I have benefited from the help of R. Prodi and A. Steinherr. Prodi allowed me to use parts of his findings on problems related to those of my research, while A. Steinherr prepared a background study on the economic consequences of the oil crisis for West Germany, from which I obtained much of my data and inferences about German economic policy during this period. I am also indebted to P. De Wolff, Edward R. Fried, and Charles L. Schultze, who kindly discussed and criticized an early version of this study. Bert G. Hickman, Jean Waelbroeck, H. Johnson, and L. Klein provided useful material and comments, which were essential in developing my research. Given the rapidly changing economic scene during and following the oil crisis, it may be of use to the reader to know that the original version of this study was prepared in September–October 1974, while the final version was written in January 1975.

The Impact on Demand and Output during the Adjustment Period

The most important effect of higher oil prices on Western Europe was the decrease in aggregate demand to which they led. The higher oil bill paid by Western European consumers to members of the Organization of Petroleum Exporting Countries (OPEC) can be thought of as an excise tax levied by the latter on the former. Initially, only part of the proceeds of this tax have been respent by member countries of OPEC on imports. Consumers, paying more for oil products, have had less available for other purchases, and sales lost by industries producing consumer goods have not been offset by increased exports to OPEC. The opposite side of the financial surplus accumulated by OPEC, therefore, has been a reduction in aggregate demand in the major industrial nations. There has been a twofold impact on aggregate demand in Western Europe: the direct impact, as outlined above, and an indirect impact from the drop in its exports to other nations whose economies were being affected similarly by the oil price increase.

The Basic Approach

In order to compute the end result of these effects, which are typical of the transfer process, it would be necessary to use a complete model of the world economy. In the absence of such a model, a simpler analytic framework must be used which nevertheless captures the major relationships among the economies of the various nations. The results will not be completely satisfactory, but they should have the virtue of allowing a relatively easy understanding of the transfer process and indicating what might be the maximum impact of the oil crisis on Western European income. Obtaining an upper boundary for the deflationary effect of the oil price increase on Western Europe can be useful in assessing its importance in bringing on the economic slowdown that has occurred in most Western European economies and in evaluating the demand-management policies adopted in response.

Our analysis of the short-run impact of oil price increases on Western Europe proceeds as follows: *First,* we estimate the initial "shock" effect of higher oil prices as equal to the increase in Western European import

Table 3-1. *Initial Shocks: Changes in Exogenous Variables Assumed in Model Simulations*

Billions of 1974 dollars

	Western Europe			United States			Japan		
	Impact of increase in oil-import bill[a]	Change in exports[b]		Impact of increase in oil-import bill[a]	Change in exports[b]		Impact of increase in oil-import bill[a]	Change in exports[b]	
Year		to OPEC (+)	to other developing countries		to OPEC (+)	to other developing countries		to OPEC (+)	to other developing countries
1974	−34.2	5.8	−0.7	−14.4	2.5	−0.5	−13.5	1.7	−0.4
1975	−28.3	6.7	−1.5	− 8.3	2.9	−0.9	−12.4	2.0	−0.7
1976	−25.1	10.1	−2.4	− 8.3	4.4	−1.4	−11.9	2.9	−1.1
1977	−20.6	13.9	−3.2	−13.5	6.1	−1.9	−11.3	4.1	−1.5

a. Difference between the oil import bill as projected with a pre-embargo ($2.75) world price and the actual and projected oil import bill under a $10 world price. The numbers are somewhat different from those projected in Chapter 1, which incorporate later revisions. An increase in the oil-import bill is shown as a negative initial shock.

b. See Chapter 1 for a discussion of the basis for these assumptions.

costs for oil plus the loss of exports to the nonoil developing countries, less increased Western European revenues from exports to oil-exporting countries. *Second,* to the net change in aggregate demand resulting from these three changes are applied multipliers, which translate the initial shock into the ultimate effect on aggregate demand and gross national product (GNP). Two sets of multipliers are needed: "own multipliers," which show the direct effect on Western Europe of the initial drop in aggregate demand, and "cross multipliers," which show the indirect effect through international trade, as falling GNP in one's own country reduces the GNP of other countries and vice versa. *Third,* the results of these calculations are then presented as deviations from the path of growth which Western Europe might have been expected to take in the absence of the sharp jump in oil prices.

The Initial Shock to Aggregate Demand

The assumed initial net changes in aggregate demand of Western Europe, the United States, and Japan attributable to higher oil imports, lower exports to nonoil developing countries, and higher exports to members of OPEC are given in Table 3-1. They are based on estimates supplied, as a framework for analysis, by the editors of this volume.[1]

1. See Chapter 1, by Edward R. Fried and Charles L. Schultze; see also a discussion of the oil market in Chapter 7, by Joseph A. Yager and Eleanor B. Steinberg.

The additional oil-import bill paid by Western Europe, Japan, and the United States reflects the increase in oil prices offset to some extent by a reduced volume of imports. Put another way, the estimates represent the difference between an estimate of what the oil-import bill would have been with prices prevailing in mid 1973 and what it has been and is likely to be with the higher prices which actually prevailed. Ideally, these estimates should include incremental payments for other energy-product imports induced by changes in the price of oil. This trade, however, is disregarded in my analysis, on the assumption that its additional magnitudes would not involve significant transfers of purchasing power between Western Europe and the other areas of the world over the next few years.

The estimates of the other two components of the initial shock—increases in exports to member countries of OPEC and decreases in exports to nonoil developing countries—are explained in Chapter 1 of this volume.

The Multipliers

The net change in aggregate demand resulting from the sum of the three autonomous changes is considered equivalent to a reduction in the autonomous component of aggregate demand brought about by government expenditures.[2] Multipliers that were originally estimated for changes in government expenditures are here applied to the net autonomous change in the trade account resulting from the oil price increase.[3] This assumption is probably acceptable for changes in exports to OPEC members and nonoil-exporting developing countries, since these constitute direct changes in aggregate demand on the same ground as do changes in government expenditures. On the other hand, the change in import value due to the higher oil bill may, like an indirect tax, be initially paid partly out of past and current savings. Thus, the assumption that it directly reduces the flow of aggregate demand by 100 percent, as a decrease in government demand would do, results in an upward-biased estimate of its deflationary impact.

2. This net change in aggregate demand, estimated by Fried and Schultze and provided as input to this study, has been incorporated into Table 3-1.

3. While incomes and therefore, implicitly, imports of Western Europe, the United States, and Japan are endogenous to the model used in this study, their exports to members of OPEC and nonoil-exporting developing countries are taken exogenously at the levels computed by Fried and Schultze.

Table 3-2. *International Elasticity Multipliers*

Percentage change in GNP of country in column resulting from a shock equal to 1 percent of GNP of country in row

Country or region	Western Europe			United States			Japan		
	First year	Second year	Third year	First year	Second year	Third year	First year	Second year	Third year
Western Europe	1.38	1.79	2.07	0.10	0.36	0.76	0.10	0.30	0.50
United States	0.06	0.16	0.33	1.18	1.87	2.58	0.13	0.27	0.40
Japan	0.01	0.03	0.05	0.02	0.04	0.06	1.18	1.50	1.50

Source: Bert G. Hickman, "International Transmission of Economic Fluctuations and Inflation," Memorandum No. 174 of the Center for Research in Economic Growth, Stanford University, August 1974. For an explanation of the derivation of the multipliers for Western Europe, see text and the appendix to this chapter.

The multipliers are those computed by Hickman from models in project LINK,[4] or are based on them. They are three-year dynamic multipliers; after the third year it is assumed that the multipliers remain stable at the level reached in Hickman's computations (which are available only for three years). His multipliers are both own and cross multipliers (in elasticity form, to do away with the national-dimension problem). Hickman gives multipliers for Austria, Belgium, France, Germany, Italy, Sweden, and the United Kingdom among the countries of Western Europe. In order to compute aggregate multipliers for Western Europe I have used weighted arithmetic means of the own and cross multipliers of these seven countries, and I have assumed that these would also be representative of the remaining countries of Western Europe.[5] (The appendix to this chapter elaborates the conditions under which it is proper to aggregate individual country multipliers by the use of weighted means.)

Table 3-2 shows the derived own and cross multipliers for Western Europe; those for the United States and Japan were taken directly from Hickman's tables. A few comments are in order concerning both the own

4. See Bert G. Hickman, "International Transmission of Economic Fluctuations and Inflation," Memorandum No. 174 of the Center for Research in Economic Growth, Stanford University, August 1974. LINK is a system of co-ordinated econometric models of various countries and regions covering the entire world. It is directed and centralized by Lawrence Klein at the University of Pennsylvania.

5. Western Europe is here defined as the aggregate of European countries that are members of the Organisation for Economic Co-operation and Development (OECD) —Austria, Belgium, Denmark, Finland, France, Germany, Greece, Iceland, Ireland, Italy, Luxembourg, the Netherlands, Norway, Portugal, Spain, Sweden, Switzerland, Turkey, and the United Kingdom.

and cross multipliers of Western Europe. It may be surprising to find that the United States is more sensitive to changes in aggregate demand in Western Europe than the reverse. Part of this result can, however, be explained.

The cross multipliers combine two effects: a decline in aggregate demand in Western Europe reduces United States exports, and this decrease in United States output is in turn subject to the own multiplier of the United States itself. Since the internal multiplier of the United States is larger than that of Western Europe (2.58 as opposed to 2.07 in the third year) the ultimate effect on the United States of a decrease in the aggregate demand of Western Europe would be larger than its counterpart, even if the initial trade effects in the two areas were the same.[6] In addition, there is evidence that the income elasticity of imports in Western Europe is significantly higher than in the United States (or Japan)—perhaps 1.5 for Western Europe as compared to 1.2 for the United States.[7] As a consequence, a decline in GNP in Western Europe would have a greater impact on the exports of Western Europe's major trading partners than a similar fall in their GNP would have on Western Europe.

These justifications notwithstanding, the possibility remains that the simplifying assumptions used in deriving the formulas for multipliers for Western Europe may have introduced an aggregation bias. This bias relates only to the own multipliers of Western Europe and to the multipliers measuring the effect on other countries' income of an exogenous change in aggregate demand in Western Europe. Thus, while there is no

6. One reason that the own multiplier is larger for the United States than for Europe arises from the monetary sector of the LINK model. The Hickman multipliers, derived from the LINK model, are not pure Keynesian multipliers, but are the results of changes in income accompanied by increases in interest rates when expansion of aggregate demand (as represented by a rightward shift of the *IS* curve) hits against a less than perfectly elastic *LM* curve. When there is more than one monetary authority, as there is in Europe (contrary to the situation in the United States), the expansion of the aggregate *IS* curve hits against nationally controlled *LM* curves, and the resulting increase in national interest rates may very well constitute a stronger brake on expansion of real income than if there were only one monetary authority. In other words, it seems reasonable to expect that the aggregate *LM* curve for Western Europe is less elastic than the *LM* curve for the United States, particularly considering that exchange controls or changes in exchange rates in individual countries of Western Europe may add some leverage to their national monetary policies. While this argument would require more theoretical and empirical analysis to be proved in fact, it suggests again a lower sensitivity of the economy of Western Europe to the United States than the reverse.

7. See footnote 14, p. 117.

reason to think that the multipliers in the second and third rows of Table 3-2 are biased, there may be bias in the multipliers in the first row along all three colunms. An indirect indication of the direction of the bias may be obtained by looking at the columns for Japan, along the first and second row. In 1973 Japan's exports to the United States were about 46 percent larger than her exports to Western Europe, yet the sensitivity of Japan to Western Europe is, according to Table 3-2, somewhat greater than its sensitivity to the United States. It is therefore likely that there is a similar upward bias in United States sensitivity to Western Europe, as well as in that of Western Europe to herself, in the figures of Table 3-2. It is difficult to assess the magnitude of the error, but it clearly goes in the direction of magnifying the deflationary impact of higher oil prices, which is consistent with the aim of obtaining an upper boundary of this impact.

The multipliers of Table 3-2 are applied to the net reductions in aggregate demand of Western Europe, the United States, and Japan that are attributable to higher oil-import values plus lower exports to nonoil developing countries less higher exports to members of OPEC, as given by the editors in Chapter 1.

Results

Combining the initial shocks estimated in Table 3-1 with the multipliers in Table 3-2 suggests that the increase in oil prices to approximately $10 per barrel reduced Western Europe's GNP in 1974 by about 2.7 percent below the level it would otherwise have reached. Assuming continuation of the $10 price, the deflationary effect will rise moderately over the next two years, reaching 2.9 percent in 1975 and then tapering off to 1.8 percent in 1977.

Table 3-3 presents the components of the total deflationary effect in Western Europe for the years 1974–77.[8] The percentages in the columns

8. In order to make the estimates for Western Europe, it was also necessary to make estimates for the United States and Japan, in order to calculate the trade effects. Precisely the same procedures were followed for those countries as for Western Europe, using the multipliers from Table 3-2. The percentage reduction in GNP below growth path for the four years beginning in 1974 were: *United States,* −1.2, −1.8, −2.7, −1.7; *Japan,* −3.4, −4.2, −4.0, −3.2. In both instances these are lower than the estimates given by George L. Perry and Tsunehiko Watanabe in Chapters 2 and 4 of this volume. Had their estimates been used the trade effects on Western Europe would have been somewhat larger, but the difference in the estimate of the impact on Western Europe would have been slight.

Table 3-3. *Estimated Effect of Higher Oil Prices on GNP in Western Europe*

Percentage change, assuming oil price of $10 and no additional energy investment

	Estimated in this chapter				Other estimates	
Year	Own effect	U.S. effect	Japan effect	Total	LINK[a]	Pekonen and Waelbroeck[b]
1974	−2.57	−0.05	−0.03	−2.66	−0.40	−1.62
1975	−2.72	−0.11	−0.07	−2.89	−1.78	...
1976	−2.51	−0.21	−0.11	−2.82	−2.49	−2.46
1977	−1.58	−0.14	−0.09	−1.81	...	...
1978	...	...	...	...	...	−1.93

Source: See text.

a. Press release, *Project LINK*, November 1974.

b. Kari Pekonen and Jean Waelbroeck, "The Impact of the Oil Crisis on the Growth of GNP in Developed Countries," International Bank for Reconstruction and Development, unpublished office memorandum, 27 September 1974.

of Table 3-2 should not, of course, be added together; they give for each year the percentage by which the growth path with the oil price increase would stay below the growth path without the oil price increase.

Unfortunately, one feature of a model such as the one used for these computations is that the growth path of GNP from which the deflationary effect is measured has to be fixed in advance. Thus, only price-determined changes in exports and imports should be included as exogenous changes in aggregate demand. However, the figures in the editors' base projections do take into account oil import reductions stemming from reductions in the GNP of Western Europe, the United States, and Japan. This introduces a downward bias in the estimate of what the deflationary effect of the oil price increase would be under conditions of full employment and target growth.[9] On the other hand, this bias may be offset by the upward biases identified above, so that the picture presented in Table 3-3 may be not too far from what the true results of the transfer mechanism would be if there were no interference from active policy changes.

It should be emphasized that in both the oil and nonoil cases, the paths compared are long-run paths in which superimposed business cycles are not taken into account. The only cycle that appears is that for which the oil crisis itself is responsible. This is why the results are not strictly com-

9. The paths are assumed to evolve at annual rates of 4.5, 4.3, and 6.0 percent for Western Europe, the United States, and Japan respectively.

parable with results obtained on the basis of complete cyclical models, such as those used in project LINK. However, the results obtained on the basis of provisional experiments performed with the LINK models are reported in Column 5. In addition, Column 6 reports results obtained by Pekonen and Waelbroeck[10] on the basis of a short-cut model which is essentially similar to the one used here and which is likewise based on multipliers derived partly from Hickman's exercise and partly from exercises performed with a model of the European Community (EC) developed at the University of Brussels. Thus, Pekonen and Waelbroeck's results are comparable to mine. Indeed, they show that my results tend to be upward-biased, but they are close enough to mine to reinforce their credibility. In my estimates the depressing impact reaches its maximum in 1975. While this is partly due to the hypothesis of multipliers leveling at their third-year value, a peak is also reached between 1976 and 1977 in Pekonen and Waelbroeck's results, which are based on multipliers that go beyond their third-year level.

Considering the peak impact, reached in 1975 in my results, it appears that the own effect of larger net imports by Western Europe, in its trade with OPEC and the developing countries, contributes 94 percent of the total Western European decrease. The recession in the United States (in its first and second rounds) contributes 4 percent, and Japan contributes the remaining 2 percent. The LINK forecasts for Western Europe are, by 1976, in substantial agreement with my computations and with those by Pekonen and Waelbroeck, discounting the fact that they are not strictly comparable with our two studies, since both the control and the nonoil solutions of the LINK models forecast cyclically perturbed economies.

The Offsetting Effect of Higher Energy Investments

I have performed a second exercise along the same lines, in which it is assumed that the price of oil stays at the $10 level but in which the extra net investment in oil-importing countries for the purpose of expanding the energy-producing sectors is taken into account.

10. Kari Pekonen and Jean Waelbroeck, "The Impact of the Oil Crisis on Growth of GNP in Developed Countries," International Bank for Reconstruction and Development (IBRD, the World Bank), unpublished office memorandum, 27 September 1974.

It is extremely difficult to obtain reliable forecasts for net additional energy investments that may be induced by the increase in the price of oil and oil substitutes. A rough idea may be obtained by considering the figures provided in a recent report of the Organisation for Economic Co-operation and Development (OECD).[11] One interesting chapter in the OECD report attempts to measure the welfare costs of higher oil prices.[12] A by-product of this exercise is a measure of the total additional revenues of domestic energy producers, which is taken to represent the additional amounts paid for salaries, for rents and profits, and for the cost of materials employed as inputs in order to expand production. (The total of these revenues is equal to the sum of Areas A and B in Figure 3-1.) Only a fraction of all this represents current net investment expenditures for producing additional energy. This total thus gives an upper bound to the magnitude we are looking for. According to the OECD report, in 1980, assuming a price of oil equal to $10.80 in 1974 dollars, the amount of the additional revenues of domestic producers of energy would be $35 billion for Western Europe—that is, about 2.4 percent of Western Europe's trend GNP in 1980.

This is certainly a high figure compared to the deflationary effects identified in the two preceding sections. Only a small fraction of this, however, represents a net increase in investment for the economy as a whole. Much of the rest constitutes the value of resources that are reallocated to the energy sector and that would otherwise have been employed in other sectors. On the whole, therefore, it is extremely doubtful that the extra investment called forth by higher oil prices will be of such great significance.

In an unpublished study by the research staff of the World Bank,[13] an estimate of total investment in production of new energy sources in all OECD countries is derived from an econometric model, assuming a $10 price of oil. Allocating this total figure for all OECD countries among the United States, Japan, and Western Europe on the basis of their extra energy production costs as given in the OECD report referred to above

11. OECD, *Energy Prospects to 1985: An Assessment of Long Term Energy Developments and Related Policies,* A Report by the Secretary-General, 2 vols. (Paris: Organisation for Economic Co-operation and Development, 1974), Vol. 1.
12. These costs, and the OECD contribution toward measuring them, will be discussed later in this chapter.
13. Jean Waelbroeck and others, "Impact of high oil prices on OECD growth," IBRD internal staff study, 1974.

Figure 3-1. *The Effect of Increased Domestic Production of Energy on Production Costs and Rents*

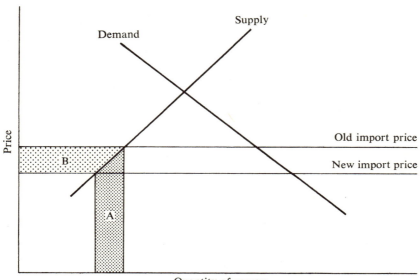

and converting the results into expenditures at 1974 prices, we obtain the estimates given in Table 3-4 for extra investment in energy sectors.

These figures for extra aggregate demand are then subtracted from the net deflationary change of aggregate demand (of the United States, Japan, and Western Europe) attributable to higher oil payments, minus higher exports to OPEC, plus lower exports to developing countries, and the multiplier analysis is performed again. The results thus obtained are presented in Table 3-5.

Comparing these results with the estimates presented in Table 3-3, we see that by 1976 extra energy investment reduces the deflationary effect of higher oil prices in a significant way for Western Europe (−1.30 instead of −2.82 percent), and eliminates the depressing impact by 1977.

In conclusion, taking my results for the case of no extra investment and the case of extra investment (both assuming a $10 price) together with Pekonen and Waelbroeck's results, and concentrating attention on Western Europe, we may safely state that the deflationary impact of higher oil prices, if left to play through multiplier effects without policy intervention,

Table 3-4. *Additional Annual Investment in Energy-Producing Sectors Attributable to Higher Oil Prices*

Billions of 1974 dollars; oil price of $10 a barrel assumed

Year	United States	Western Europe	Japan
1974	3.52	3.02	0.12
1975	8.50	7.28	0.29
1976	16.24	13.92	0.56
1977	16.89	14.48	0.58

Source: Jean Waelbroeck and others, "Impact of High Oil Prices on OECD Growth" (International Bank for Reconstruction and Development, 1974; processed).

would produce a maximum trough which should be somewhere between −2.4 and −2.9 percent below the income path that might have been followed in the absence of the oil price increase. This trough would be near the lower boundary (−2.4) and would occur in 1974 in the case of extra energy investment and would be near the upper boundary (−2.9 percent) and would be postponed for a year—to 1975—in the case of no net extra investment attributable to energy development.

Even if the incremental investment figures that we have used are correct, however, they may contribute less to an expansion in aggregate demand than our calculations suggest. There has been a significant rise in the price of coal produced in Western Europe, and a large part of this rise represents a transfer of income from consumers to rents received by coal-producing companies. To the extent that these rents are not transferred back to consumers (through increased dividends paid by coal companies, for example) they bring about an increase in savings. Only the increment in energy investment in excess of this additional saving will result in an increase in aggregate demand to offset the demand-depressing effects calculated in earlier sections of this chapter.

Table 3-5. *The Effect of Higher ($10) Oil Prices, Offset by Larger Net Energy Investment, on GNP of Western Europe*

Percent

Year	Own effect	U.S. effect	Japan effect	Total
1974	−2.30	−0.04	−0.03	−2.38
1975	−2.02	−0.05	−0.07	−2.14
1976	−1.15	−0.04	−0.11	−1.30
1977	−0.00	+0.12	−0.09	+0.03

Source: See text.

The Effects on the Trade Account

Our estimates of the effect of higher oil prices on aggregate demand took into account the interactions among the three major regions. These interactions operate chiefly through the trade account, as changes in prices and income in one region affect the exports and imports of the other regions. Their effect on GNP is summarized in the cross multipliers. But the LINK cross multipliers, used for that purpose, are not in a form which allows a direct calculation of the oil-induced changes in the trade accounts of each region. In this section we attempt to estimate those changes.

For the $10 oil case (with no extra energy investment), we apply to the changes in GNP roughly estimated income elasticities of demand, as shown below:[14]

Country or region	Short-run (first year)	Long-run (second year and after)
Western Europe	0.7	1.5
United States	0.6	1.2
Japan	0.5	1.0

The changes in total import demand induced by changes in income in the United States and Japan are then allocated to Western Europe on the hypothesis of constant shares (based on 1972–73 averages). In addition, increased exports to OPEC and lower exports to nonoil-exporting developing countries are added to obtain the total change in Western European trade accounts attributable to the oil price increase. As is shown in Table 3-6, the increased Western European oil bill of $34.2 billion in 1974 leads to an additional trade deficit of $27.2 billion when these various effects are taken into account. The Western European trade account with nations other than members of OPEC improves somewhat—by $1.0 billion in 1974—principally because of the higher income elasticity of imports assumed for Western Europe than for Japan and the United States.

Over the three succeeding years—1975 through 1977—the incremental effect of the oil price increase on the balance of trade falls sharply,

14. These elasticities are based on rough comparison of estimates obtained by Hendrik Houthakker and Stephen Magee, by the OECD model, and by models in Project LINK. For a summary table on the first two sets of estimates, see *The Economist,* 18 January 1975, p. 69. For the third set of estimates, see Giorgio Basevi, "Commodity Trade Equations in Project LINK," in R. J. Ball (ed.), *The International Linkage of National Economic Models* (Amsterdam: North-Holland Publishing Company; New York: American Elsevier Publishing Company, 1973).

Table 3-6. *Changes in the Western European Trade Account Attributable to Higher ($10) Oil Prices*[a]

Billions of 1974 dollars

		Trade with OPEC			Trade with rest of world			
Year	Total impact	Total	Larger oil bill[b]	Higher exports to OPEC[b]	Total	Exports to nonoil developing countries[b]	Exports to United States and Japan	Lower Western European imports
1974	−27.4	−28.4	−34.2	+5.8	+1.0	−0.7	−0.2	+1.9
1975	−19.2	−21.6	−28.3	+6.7	+2.4	−1.5	−0.6	+4.5
1976	−13.3	−15.0	−25.1	+10.1	+1.7	−2.4	−0.8	+4.9
1977	−6.5	−6.7	−20.6	+13.9	+0.2	−3.2	−0.8	+4.2

Source: See text.

a. Estimates are based on the assumption of no additional energy investment.

b. These estimates are somewhat different from those projected in Chapter 1, in which later revisions were incorporated.

from the −$27.4 billion of 1974 to −$6.5 billion in 1977. Most of this improvement results from the fall in the volume of oil imports and the rise in exports to members of OPEC. The Western European balance of trade with the non-OPEC world also improves somewhat in 1975 and 1976, again because of Europe's relatively high income elasticity of imports.

Effects on Import and Export Prices

The increase in the price of oil brings about changes in the prices of other commodities imported into Western Europe and in the prices of commodities that compete with Western Europe's exports. In addition, it changes the level of activity and associated conditions of import demand in the markets to which Western Europe exports. Besides these effects, the oil price increase will influence financial variables that are exogenous to Western Europe, such as interest rates in the United States, OPEC investments in the United States, or aid to nonoil-exporting developing countries. International financial effects of the oil crisis, however, will not receive much attention in this chapter, since they are the subject of another chapter.[15]

The increase directly induced by the higher oil price in the prices of

15. See Chapter 6, by John Williamson.

other primary products imported by Western Europe (those classified under the Standard International Trade Classification [SITC] groups 1, 2, and 4) is mainly due, in the case of agricultural products, to the increased price of energy and fertilizers; in the case of rubber, to the increased price of oil-based synthetic substitutes; and in the case of ores and metals, to the increased price of energy. According to computations performed on the basis of the models in the LINK project, the unit values of primary products in total world trade would have shown, with and without the oil price increase, the following percentage rates of change:

	1975	1976
With oil price increase	7.06	5.11
Without oil price increase	4.43	2.55

The inference is that, by 1976, the prices of internationally traded primary products would have stabilized at a level 5.1 percent lower than the level forecast as a result of the oil crisis.[16]

The LINK estimates, of course, do not simply measure the *direct* effect of higher oil prices on the prices of other primary products, but also the indirect effects arising out of changes in overall levels of world activity. Since without the oil crisis the rate of world activity from 1974 to 1976 would have been higher, the rate of 5.1 percent is probably a downward-biased estimate of the impact of oil on the prices of other primary products. It is difficult to believe, however, that the direct effect of higher oil prices on other import prices would be more than twice this figure. Thus the most likely estimate may be somewhere between 5 and 10 percent.

The oil-induced increase in the prices of manufactured products imported into Western Europe is of a similar order of magnitude. The LINK estimates give the following percentage rates of change for unit values of manufactured products in total world trade:

	1974	1975	1976
With oil price increase	19.22	7.62	4.77
Without oil price increase	12.70	6.21	3.85

In this case, by 1976 the import prices of commodities classified in SITC groups 5–9 would have been at a level 8.1 percent lower than they are forecast to be with the oil effect. As before, these estimates are likely to represent the lower limit, but in this instance the range of uncertainty is

16. These estimates understate the induced impact of higher oil prices on the prices of other primary products, since the estimates only take account of the impact from 1974 onward, whereas there was some impact in 1974 itself.

narrower, since the prices of manufactured products fluctuate less with the business cycle than do those of agricultural products, rubber, and ores and metals.

Indirect estimates of probable value can also be obtained from input-output computations, since exports of industrial countries for which these computations are available make up a large part of the trade in chemicals, manufactured goods, and other items included in SITC 5–9. On the basis of mechanical input-output transmission of higher oil prices (percentage rates of change), the deflators of commodity exports should increase as follows:[17]

Germany	1.6
Belgium	6.0
France	5.6

On the other hand, estimates based on the LINK models (and thus incorporating more than input-output mechanical transmission) give the following increases for export price indices during 1974 in relation to the control solution without the oil price increase:

Germany	1.1
Italy	4.8
United Kingdom	7.0

Conceptually the input-output results should give lower results than LINK for Western European export prices. Exports of those items included in SITC 5–9 by Western Europe are generally of a high degree of manufacture and are probably more intensive in labor and capital than in energy. LINK includes the effects of higher oil prices on wages and other costs of inflation, while input-output computations leave the price of value-added components unchanged. We may thus retain 8 percent as a lower boundary for the oil-induced price increase of manufactured products imported by Western Europe, with an upper boundary not far above.

Limitations of the Model and of Its Results

It would be a mistake to place a high degree of faith in the estimates obtained above. As already explained, they suffer from technical biases, to which, however, a direction and probable magnitude can be attributed in

17. For Germany and Belgium, see OECD, *Country Report,* 1974; for France, see C. Gabet, G. Honoré, and F. Houssin, "Les répercussions mécaniques des hausses des prix énergétique," in *Economie et statistique,* 56 (May 1974), pp. 45–50.

most instances. The errors are essentially of three types: aggregation biases, emphasized earlier with respect to Western European multipliers; the assumed linearity of the economic relations, used implicitly in applying Hickman's multipliers to a period outside the sample on which they were obtained; and leveling of the multiplier effects in a somewhat arbitrary way at the third year.

More important, however, may be the failure of the analysis to take into account some theoretical and empirical characteristics of the oil crisis and of its effects on the economy of Western Europe. A first element of empirical refinement would be consideration of the additional change in Western Europe's terms of trade attributable to the increase in the prices of imports other than oil—an increase brought about by higher oil prices—and the increase in the prices of Western Europe's exports for which higher energy costs are responsible. These effects were discussed earlier, and it was shown that the oil-induced increase in import prices (other than that of oil) might range from 5 to 10 percent, while the oil-induced increase in overall export prices might range from 8 to 10 percent; there is thus a small possible improvement in Western Europe's terms of trade (to be subtracted, of course, from the large deterioration accounted for by the increase in the price of oil). Because it is comparatively small, the offsetting movement does not significantly change the picture drawn above of the transfer mechanism directly implied by the oil crisis; its direction, however, indicates that the depressive effects identified earlier may be somewhat overestimated.

A second set of problems that has hitherto been avoided or swept under the carpet in our macroeconomic analysis concerns sectoral and geographical disaggregation. In a way, the geographical aspect was touched upon tangentially in the discussion of aggregation biases in the multipliers for Western Europe. But the problem is more fundamental. In fact, the overall picture given there for Western Europe hides widely differing patterns of reaction among the countries of Western Europe, both with respect to their income paths and to the shocks and readjustments in their trade account. It was, however, impossible to derive the estimates on a more detailed country disaggregation, both because of lack of data and because of the decision to rely on a simplified and manually computable model, which would have become excessively cumbersome to operate if the number of areas had been increased beyond those considered in this exercise. On the other hand, some discussion of the problems of specific countries of Western Europe will be offered in the policy section of this chapter.

Not only is there insufficient national disaggregation within Western Europe, sectoral disaggregation is also lacking in the analysis of the transfer mechanism. The changes in domestic prices induced by the oil price increase affect the whole structure of sectoral supply and demand and therefore affect exports and imports of particular products. The short-run and long-run reactions in the trade account will be different. Waiting for structural readjustments toward a new pattern of international trade specialization, a country—or a group of countries such as Western Europe —may find itself in need of importing goods that, once the reallocation of production is completed, it might then be able to export. It is thus very hard to estimate in the aggregate either the magnitude or the timing of these effects upon the trade account; for the compelling, albeit unsatisfactory, reason of ignorance, therefore, they have been left out of this analysis.

A third set of empirical consequences left out of the picture involves trade in services that accompany, or may in any case be influenced by, the flow of trade and its oil-induced changes. The focus of this study, however, is essentially on the trade account, not the overall current account. It remains true, on the other hand, that important reactions in the trade in services may be considered, at least conceptually, in order to give a clearer picture of the overall economic consequences for Western Europe and see whether considering them makes the outlook worse or better.

The services that will clearly be affected by the oil crisis are profit revenues from oil companies; insurance, banking and other professional and consulting services; transportation services; technical services of many kinds, such as design and engineering services and the training of personnel; and interest payments on the foreign debt accumulated to finance current-account deficits.

Roughly speaking, Western Europe should suffer less from the decrease in oil revenues of multinational companies than the United States; Japan should suffer still less. Western Europe and the United States may gain from the financial intermediation and other consulting services that accompany the need to invest and recycle the financial resources initially transferred to OPEC. All three zones should participate heavily in the growing export of technical services to the oil-producing countries. A more difficult question, but one that is critical because of its repercussion on Japan's export of tankers, is the effect of higher oil prices on the oil trade and overall world trade. Freight rates are hit by a decline in world trade, and if the decline starts from a fall in the volume of the oil trade, the rates of oil tankers are hit particularly hard.

On most counts in the area of services, indications are that Western Europe may come out the best of any of the three zones, despite the fact that some financial and consulting services earned in Western Europe may eventually be transferred in the form of profits to multinational corporations based in the United States. With respect to interest payments on foreign debt, however, Western Europe may come out very badly, with current accounts much weakened by the need to provide for large interest payments on debts incurred to finance their trade-account deficits. This question will be taken up again later, in a discussion of the long-run welfare and policy outlook.

While most of the limitations pointed out above (with the exception of sectoral and geographic disaggregation) could be handled by the model presented earlier, with the addition only of more empirical knowledge, there are two other theoretical elements that escape analysis by means of a simple model, yet they may be of particular importance.

The way to consider the net deflationary impact of the oil price increase, as already explained, was to treat the larger imports of oil less the larger exports to OPEC plus the smaller exports to the developing countries as an exogenous change in aggregate demand. An alternative way, theoretically more satisfactory, would be to introduce the deflationary impact of the oil price increase by means of its effect on the final prices of each component of aggregate demand.[18] The advantage of this procedure is that it feeds the deflationary impact into demand through the reduction of real disposable income, on which consumption is based, and through the increase in the price of investment, government, and export expenditures separately. Thus it allows for a more disaggregated view of the multiplier effects of the oil price increase. Moreover, if import propensities are influenced, not only by the overall level of aggregate demand, but also by its composition in terms of consumption, investment, government, and exports, a model capable of demonstrating all these effects would be much more informative. For Western Europe as a whole a very large model would have been required.

18. This alternative is adopted by George L. Perry in Chapter 2.

A second aspect of this problem arises from looking at the flow of income, not from the expenditure side, but from the side of its distribution. The rise in the price of oil may alter substantially the distribution of income between wages and profits, particularly between normal profits and extra profits or rents, which of course are typically represented by extra profits in the domestic oil-producing sectors. Insofar as such a redistribution from wages to profits, and from normal to extra profits (or rents), occurs, it may substantially change the normal pattern of reaction of the economy—that is, the normal propensities to hoard (not to spend) and to spend for consumption, for investment, and for imports.[19] In particular, at least in the short and medium run, there is the likelihood that an increase in extra profits many increase the propensity to hoard, thereby introducing a further deflationary effect into the economy. Given actual and potential capacities for the production of energy, it is also likely that this danger will be higher in the United States and in the northern countries of Western Europe, while it may be much lower or even entirely absent in the southern countries of Western Europe and in Japan. Thus, the differences in deflationary impacts traced earlier for Western Europe, the United States, and Japan may be reduced further.[20]

Further reactions to the redistribution of income between wages, profits, and rents—particularly interesting with respect to the rate of economic growth—will be discussed later. It may be useful to anticipate at this point, however, that if the higher prices for domestic and imported energy were to cause a reduction in profits and investment in other industries, and if the import content of investment should be higher than that of consumption, there would follow a relative improvement in the current account, at least in the initial years. The reduction of potential output that would then result from the lower investment expenditures, by decreasing the growth of capacity in subsequent years, would increase imports and reduce exports at that time. Thus, besides a lower growth path, a reduction of investment on account of lower profits would introduce into the trade account a swing contrary to that observed in the transfer mechanism—that is, a relatively better performance in early years, followed by a relatively poorer one in later years.

19. Indeed, as will be discussed later, normal profits will probably fall, to the advantage of larger rents.
20. President Ford's proposal for a tax on profits of domestic oil producers may be interpreted in part as an attempt to take away extra profits from a sector that may have a low propensity to spend them and redistribute them to persons or businesses with higher propensities to spend.

Policy Reactions in Western Europe: An Assessment

In the light of the above analysis, how well did the economic policies of Western European governments deal with the problems raised by the oil crisis? Economic policy will be broadly divided into actions aimed at the initial supply impact of restricted availability of oil during the winter of 1973–74 and the concomitant (and later more important) interventions the aim of which was to control the impact of higher oil prices on inflation and on the level of aggregate demand and production.

Policy Aimed at Controlling the Allocation of Oil and the Effects on Supply[21]

The progressive reduction of production decided upon by most Arab countries between October 1973 and March 1974 did not produce dramatic reductions in the availability of oil for the European countries, because of larger imports from non-Arab countries (especially Iran and Nigeria), reduction in oil consumption as a result of the price increase, and, to a lesser degree, energy-saving measures adopted by most Western European countries. Indirect support for this statement is provided by data on the level of stocks, which never fell below an amount sufficient for eighty days' normal consumption. Even in the countries that have suffered most from the oil crisis, the stocks have always been normal: in Italy, where the situation was probably the worst, stocks increased by 23 percent, or twenty days' normal consumption.

The magnitude of the crisis, therefore, as measured by restrictions on the oil supply, was not dramatic, and it certainly was not comparable to the reductions experienced during the Middle East crises of 1956 and 1967. Nonetheless, Western Europe was also significantly affected, although oil was seldom actually in short supply. The degree of European industry's dependence upon imported oil is in fact much larger than that of the United States, not only in terms of the absolute level of imports, but also with respect to the pattern of consumption. Twenty-one percent of U.S. oil consumption is used for industry, including feedstocks for petrochemicals, compared to thirty-eight percent in Western Europe.

21. This section is based on a study by Alberto Clò and Romano Prodi on the responses of Western European governments and international organizations to the oil crisis. I am grateful to the authors for allowing me to draw on the provisional conclusions of their research.

Perhaps more important in weakening Western Europe's position and paralyzing its supply-control policies were the psychological consequences of the embargo and of the discriminations that accompanied the new oil policy adopted by the Arab countries. In Western Europe the embargo was applied only against the Netherlands—which, however, is strategic in oil trade and refining. Holland re-exports to its neighboring countries 1.2 million barrels of crude oil and 0.8 million barrels of refined products per day. Thus, to frustrate the embargo by official action would have required a policy of redistributing oil imports within Western Europe, an action that did not prove to be possible because of the fear of Arab retaliation. The Dutch government's request for solidarity on the one hand and the concern of France and England to avoid any confrontation with the Arab countries on the other hand created a vulnerable state of indecision. As a consequence, the reallocation of oil among the Western European countries was left for the most part to the international oil companies. Allocation was thus accomplished through the economic forces of markets and prices, with the result that Holland, Germany, and the United States got the oil, while other countries that did not increase prices quickly enough suffered more than these three countries did. In fact, international oil companies controlled 65 percent of European oil supplies directly— and 19 percent indirectly—and were therefore able to discriminate in favor of their branches in order to delay deliveries to their competitors and to work an effective hardship upon some countries (Italy and Belgium, for example) that did not increase the price of refined products quickly enough.[22]

The internal measures taken by Western European governments beginning in November 1973 in the face of the supply crisis were aimed at reducing the level of demand for oil and increasing its internal availability by forbidding or limiting exports of refined products. This strategy was adopted by all major net exporters of refined products—the United Kingdom, Belgium, Italy, Holland, and Spain. It is important to stress that no distinction was really made between members and nonmembers of the European Community (EC), even though these controls were against EC regulations.

22. Chapter 7 contains a more comprehensive discussion—differing somewhat from this one—of the way in which the Arab embargo and cutbacks in production affected various oil-importing countries.—Eds.

The only result of these decisions was to create important distortions of intra-European trade in oil and oil products—distortions that had negative consequences not only on the traditional importing countries (such as Germany) but also on those that traditionally exported oil products. It was, in fact, useless to control the export of refined products without having the control of imports of crude oil at the same time: clearly the international oil companies were capable of redirecting crude oil supplies away from countries that adopted controls on the export of refined products.

All European governments tried to reduce private consumption in order to alleviate the difficulties of industry. The most common decision was to reduce gasoline consumption (which amounts to only 12 percent of total oil consumption) through the prohibition of automobile travel during holidays (in Italy, Holland, Belgium, and Denmark), the reduction of automobile speed limits, and the closing of filling stations for longer periods than before. Other limitations upon heating and electricity were planned, through rationing of heating oil (in the United Kingdom, Italy, Holland, Belgium, and Germany), closing of shops and public offices, reduction in public lighting (in Belgium, Italy, and the United Kingdom), and restriction in the supply of heavy fuel oil (in Holland −15 percent, in the United Kingdom −10 percent, and in Ireland −5 percent).

Italy and the United Kingdom suffered more than other countries from the scarce supply of heavy fuel oil, but in the United Kingdom the situation was made especially serious by strikes in the mining and electrical industries, which led to the adoption of the three-day work week in commerce and industry.

It is difficult to measure the contribution of these control policies in limiting the growth of petroleum demand, because a slowing down had already been noted in Western Europe during the third quarter of 1973 (real growth of 4 percent instead of the 8 percent growth experienced during the first half of 1973), and a further decrease was forecast for the fourth quarter. All of Western Europe, moreover, experienced a mild winter.

On the whole, however, most of the reduction in oil consumption effected in Western Europe (as well as in other OECD countries) can be attributed directly to the depressing effects of higher prices (through price-elasticity effects) and indirectly through their deflationary effect on income, which came, as we shall see presently, on top of an already slump-

ing economic situation. As the situation is summarized, with some dis-
illusionment, in the OECD *Economic Outlook,*

no very determined efforts seem so far to have been made to economise on the
use of oil. An ambitious objective in this field has been announced by France,
but at the time of writing details are not available of the precise measures
envisaged to achieve it;[23] the United States has also announced an objective in
this area. In all countries there has been much discussion of oil saving but little
concrete action since the oil crisis first developed.[24]

Policies Aimed at Managing Aggregate Demand and Inflation

The oil crisis came at a critical moment in the business cycles of most
OECD countries. On the whole, 1973 had been a year of great expansion
in real production, with the growth of GNP of the seven major OECD
countries amounting to 6.5 percent and that of the entire OECD to 6.3
percent. It was apparent, however, that in most OECD countries the level
of output had come so close to capacity as to superimpose supply-de-
termined inflation on an already steep path of price acceleration that had
initially been fed by excessive demand.

In the OECD as a whole, consumer price increases accelerated from
4.7 percent in 1972 to 7.7 percent in 1973 (during 1961–71 the average
rate of inflation had been 3.7 percent). In Western Europe in particular
they accelerated from 6.7 percent in 1972 to 8.7 percent in 1973 (whereas
from 1961 to 1971 the average rate had been 4.2 percent).

It is safe to say that this high rate of inflation was essentially the result
of excessive creation of money in the years preceding and particularly of
an amazing expansion of the Eurocurrency markets. When, in the years
1971 to 1973, the fixed-exchange-rate system was gradually dismantled,
there was a decrease in the need for international reserves on the part of
central banks, while the process of international money creation continued, particularly through the Eurodollar market.

While this monetary expansion had helped certain countries to pull out
of the recession that they had experienced during 1970–71, it was so
pronounced by the end of 1973 that it had brought the OECD countries

23. The "ambitious objective" aimed at by France is not to break a ceiling of 51
billion francs on total oil imports for internal consumption in 1975.
24. OECD, *Economic Outlook,* 16 (December 1974), p. 11. This report was, of
course, published before the request made in January 1975 by President Ford for a
tax of $3 a barrel on imports of oil.

to a coincident peak in their business cycles.[25] It was inevitable that these countries, all having come so close to their output ceiling at the same time, could no longer use monetary balances for further expansion, but, particularly in the face of inflation, would channel additional money supplies into stocks of commodities, especially metals and other primary products. Thus we experienced an exceptional increase in the prices of agricultural products and other raw materials, which was of course exacerbated in the case of agricultural products by a series of crop failures.

At the beginning of 1974 inflation in consumer prices was reaching historic peaks (12.8 percent from April 1973 to April 1974 in Western Europe), while unemployment was still quite low and was even falling in Italy and France, for example, during the first quarter of 1974. This was the situation when the oil crisis made its appearance.

The initial policy reactions—restriction and reallocation of oil supplies among various countries, accompanied (and later replaced) by extremely high price increases—made the danger of inflation seem even greater[26] and made more apparent the pressure of excessive demand against exogenously restricted and sectorally disrupted supply. It was now possible to justify the adoption of sharply restrictive monetary policies (or continuation of those which had been adopted in 1973), in order to choke inflation by reducing demand to a more nearly sustainable level in relation to potential supply.

Monetary authorities, which in most countries had already moved toward restriction in 1973, kept and even intensified their tight stand, without regard for the lags previously observed in the effectiveness of monetary policy. Only Germany among the four largest countries of Western Europe began relaxing monetary policy at the end of December 1973. France kept strict limits on the growth of credit from July 1973 to July 1974, even intensifying them thereafter, at the same time making some attempt to attenuate their impact on export credit and on investment

25. See OECD, *Economic Outlook,* 14 (December 1973), p. 24, Chart B. The previous point of maximum coincidence had been reached in 1963.

26. Yet it was apparent that oil price increases were contributing only a relatively small fraction to the rise of the deflator of total domestic expenditure. For 1974, the OECD, *Economic Outlook* of July 1974 was estimating the contribution at 1.5 percentage points for the whole OECD area (as against 1.0 percentage point for which primary products other than oil were responsible) and at 2.3 percentage points for Western Europe (as against 1.7 percentage points accounted for by primary products other than oil). See OECD, *Economic Outlook,* 15 (July 1974), Table 13, p. 30.

in energy sectors. Italy, which had a looser monetary policy during 1973 than the other countries of Western Europe, made it progressively more restrictive during 1974, particularly in the spring, through the monetary impact of the special import-deposit scheme that was introduced in May. Italian monetary policy, however, began to be relaxed somewhat in the fall of 1974. In the United Kingdom monetary policy, which had been tightened considerably in December 1973, was subsequently relaxed somewhat, but the rate of expansion of the money supply remained lower than the growth rate of the GNP in current prices. In November 1974 new measures were taken with the purpose of redirecting funds toward productive investment while keeping overall monetary policy restrictive. In the smaller countries of Western Europe, monetary policy remained on the whole quite restrictive throughout 1974, particularly in Spain, Sweden, and Switzerland.

While restrictive monetary policy, influenced by preoccupation with inflation and the balance of payments, seems to have prevailed in most Western European countries, fiscal policy varied substantially from country to country. In Germany, expansionary moves were undertaken as early as December 1973. They were far from adequate, however. On the whole, an early move and the preparation of regional and sectoral programs of help were not followed during 1974 by much in the way of concrete measures to meet the need of reflating demand in a situation characterized by a healthy balance of payments and a comparatively strong anti-inflationary performance. A substantial fiscal expansion was expected to come only with the fiscal reform going into effect on 1 January 1975.

In the United Kingdom, after a mildly deflationary budget adopted in March 1974, the government intervened again in July with expansionary measures (based primarily on reductions of value-added tax, domestic rate reliefs, and larger transfers) and again in November, in order to favor investments and exports.

In France, on the other hand, fiscal policy was tightened in December 1973 and tightened further in June 1974; no change of policy toward fiscal expansion was indicated in plans for 1975.

Italy made a substantial move toward fiscal restriction in August 1974; despite rising prices, budget expenditures in nominal terms were to be held constant in 1975. Recent negotiations to increase old-age pensions would seem to put that supposition in doubt, however.

Table 3-7. *Actual and Forecast Levels of GNP in Western Europe Compared with Trend Growth of GNP Adjusted for Oil Price Increase*

Index: 1973 = 100

Year	Actual and forecast GNP[a]	GNP adjusted for oil price increase
1973	100.0	100.0
1974	102.8	101.7
1975	105.1	105.8
1976	...	110.2
1977	...	116.2

Sources: "Actual" from official statistics; "forecast" from OECD, *Economic Outlook*, 16 (December 1974); "trend adjusted for oil" from the estimates presented earlier in this chapter with the assumption of an oil price of $10 a barrel and no additional energy investment.

a. Figures for 1973 and 1974 are actual; figure for 1975 is forecast.

Policy Appraisal

The brief outlook presented above is not sufficient for an appreciation of the detailed effects of demand-management policy during the upswing that preceded the oil crisis and the downturn that began in 1973 and seems to be deepening in 1975. To what extent, in other words, has the current business cycle been provoked or intensified by rises in the price of oil on the one hand and badly tuned economic policies on the other? To answer this question, it may be useful to compare the actual or forecast path of real GNP with that resulting from the computations presented earlier in this chapter, in which a trend path was shocked by the deflationary effects of the oil price increase.

Only the deflationary effects obtained from these computations, assuming a price of $10 a barrel for oil and no net extra energy-induced investments, are presented for comparison in Table 3-7, since, as explained earlier, the comparison probably represents an upper limit to the deflationary effects of the oil price increase in the absence of offsetting policies. Any additional deflationary effect, if present, should be due to other pre-existing economic forces or to badly managed economic policies.

In Western Europe, we find that for 1974 the development of the economy looks better than it might have if the deflationary effects of the oil crisis had followed their course without active policy intervention. We may thus advance the hypothesis (which must be substantiated by deeper econometric work) that either the recession phase of the cycle (other than that attributable to the oil price increase) was of lower intensity in Western Europe or that demand-management policies may have had a softening (that is, a net expansionary) effect there.

The fact that in 1974 the actual growth path for Western Europe was above that of the oil-deflated path, while in 1975 it goes somewhat below it, suggests that the explanation lies in the stronger endogenous performance of some Western European economies rather than in more enlightened demand management. In particular, the business cycle in Italy and France, delayed in relation to that of the other major countries of the OECD, allowed these two countries, particularly France, to expand throughout most of 1974. In addition, demand-management policies in Italy may have delayed and softened the deflationary impact of the oil price increase further, not because of a decisive move in this direction, but because of the difficulty of actually enforcing a restrictive monetary stand through the first half of 1974 in the face of a government deficit that was growing unmanageably. Also, the delay in adopting a firm incomes policy in the United Kingdom, widespread indexing of wages and salaries in smaller European countries, and generous unemployment benefits in Germany and other northern countries of Western Europe have generally kept real wages from being undercut by the high rate of inflation, in contrast to what happened in the United States. Of course, this may not augur well for the future, in view of the fact that the oil tax, which must be paid by Western Europe in one way or another, has apparently fallen mainly on profits, thus decreasing investment and preparing the ground for future economic difficulties.

Thus, institutional and structural characteristics, rather than better-managed economic policies, account for the fact that until the end of 1974 Western Europe seems to have run a better course than it might have run had it been left exposed to the purely deflationary effect of higher oil prices. The other face of the coin, of course, is a much poorer performance in the trade account, particularly since deficits have been so heavily concentrated in three countries—the United Kingdom, Italy, and France.

It is difficult to suggest, on the basis of this broad and sketchy analysis, any guiding principles for the future course of demand-management policies. The question is, what can the governments of Western Europe do, besides hope for a more expansionary policy in the United States? Some elements of an answer can be inferred from the preceding discussion. In fact, insofar as a more expansionary level of aggregate demand was maintained in Western Europe through income policies (obtained de facto, or actually managed) and easier fiscal policies, while monetary policy remained restrictive, interest rates were pushed to dangerously high levels in the face of declining profits.

Thus, while aggregate demand has been maintained, its composition has been biased, somewhat toward government expenditures but principally toward consumption and away from investment. This situation must certainly be corrected soon if serious problems for the future are not to accumulate.

There are two arguments for a larger share of investment in real expenditure in Western Europe. First, the current-account deficits are causing the accumulation of debts which will ultimately have to be serviced by real transfers of income. Future losses of income will, consequently, be greater than the loss attributable to higher oil prices alone, unless the increase in debt is accompanied by increases in Western European investment for the purpose of generating higher productivity and income. Second, unless this is done in 1975–76, there is danger that too great an acceleration of investment will accompany an accelerating export demand, and excessive aggregate demand will reappear around 1977–78.

Thus, demand-management policies and income policies for 1975 and 1976 should concentrate not only (or not so much) upon maintaining the aggregate level of demand (and we may hope that this task may be accomplished to some extent by larger expenditure in the United States) but also upon increasing the incentive to investment. Accomplishment of this objective will require a more expansionary monetary policy, together with perhaps a neutral or mildly expansionary fiscal policy. Policies with respect to income and taxation must be such as to maintain or increase the level of profit, so that there will be not only more readily available credit for financing investment, but also a more favorable outlook for the profitability of investment.

Long-Run Effects and Issues

The long-run effects of the oil price increase may be defined as those that will remain once the deflationary impact has run its course and the oil-importing countries have returned to full employment.

These effects are of three types: effects on overall community welfare under static conditions (macrostatic effects); effects on the characteristics of the economic structure (micro effects); and effects on income distribution and economic growth (macrodynamic effects).

Macrostatic Effects

These effects are essentially connected with the change in the terms of trade accounted for by the oil price increase. The loss in terms of trade may be conceptually and empirically measured under the assumption that long-run equilibrium in the trade account will be re-established and that there are no major distortions in the economy. It is given by the difference between the new and the old relative price of imports times the new level of imports, plus the cost of producing domestically what would otherwise have been produced abroad and of restricting demand for importable goods.

Graphically, this is represented by the shaded area in Figure 3-2, where $m_2 m_3$ represents the new quantity of imports and w' its new price in terms of exports that are necessary to pay for them.

As shown in earlier analysis, the nonoil terms of trade do not seem to be affected significantly in a clear direction by oil-induced changes in the prices of other products. It may thus be sufficient to consider Figure 3-2 as applicable only to net energy imports in the new long-run situation of supply. The shaded area is approximately equal to

$$M \cdot dp + \frac{1}{2} \cdot dM \cdot dp$$

where M is the new long-run level of energy imports and dp is the change in the relative price of energy imports. The first element in this expression (corresponding to Area B in Figure 3-2), may be called the "terms-of-trade loss" in the narrow sense,[27] and the second element (corresponding to Areas A and C in Figure 3-2), "the inefficiency cost" (in production [A] and in consumption [C] respectively).

Computations for the terms-of-trade loss are provided in the OECD energy study[28] under the assumption of a price of $10 a barrel for oil. By 1980, this cost would amount to 2 percent of the GDP of Western Europe, considering it as applied to trade in all energy materials and not simply in oil. A provisional version of the same OECD report also provided a

27. Both elements add up to what is called, in the real theory of international trade, the terms-of-trade loss.

28. OECD, *Energy Prospects to 1985,* Vol. 1, pp. 67–68.

Figure 3-2. *The Long-Run Terms-of-Trade Loss Attributable to Higher Energy Prices*

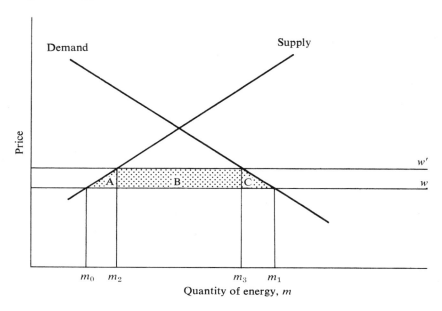

measure of the "inefficiency loss": 0.8 percent of the GDP of Western Europe in 1980. Thus, on the whole, the long-run terms-of-trade loss might be estimated to reach a maximum of 2.8 percent of the GDP of Western Europe. This is an amount which should not prove unbearable, given the long-run growth rate of GNP potentially sustainable by our economies. However, it looms large in terms of international distribution and willingness to adapt to it, if, for example, we compare it to the amount of aid given by developed countries to developing countries, which did not even in the aggregate reach the 1 percent of GNP ratio set by the United Nations as a target at the beginning of the sixties.

On the other hand, the figure is a maximum estimate, for the simple reason that this loss (which includes the long-run terms-of-trade tax to be paid to OPEC) need not be paid in full. As already mentioned, the occurrence of this loss is based on the assumption that trade will be balanced in the long run—that is, that the transfer of purchasing power implicit in the higher price of oil has been converted into a real transfer of goods through larger exports to members of OPEC. During the adjustment period analyzed earlier, this, of course, is not the case, since the trade account goes directly into deficit as a result of increased oil prices, and a few years are required for Western Europe to reabsorb this additional net deficit through the deflationary effect of higher oil prices on income (and through the expansionary effects on OPEC imports).

It is only insofar as demand management brings the economy back to full employment, at the same time allowing the additional oil-induced deficit to be reabsorbed through a transfer of real resources to OPEC, that the full long-run terms-of-trade loss will be incurred. If in the long run the members of OPEC (and Western Europe for its part) find it expedient not to reabsorb the deficit wholly, but accept a continuing inflow of OPEC funds into oil-importing countries, the long-run terms-of-trade loss will be reduced. Since this structural development in the current accounts of Western Europe and OPEC may be a reality for the long run, the 2.8 percent terms-of-trade loss measured in the OECD study is likely to represent an upper boundary to the magnitude of the actual terms-of-trade loss.

Sectoral and Microeconomic Effects

The losses, both transitional and permanent, discussed above, are macroeconomic in character. In current discussions, however, more attention is sometimes given to losses that may originate in the microeconomic structure of production and employment. The increase in the relative price of oil and other energy materials gives rise to a complex pattern of reallocation of demand and production and may therefore cause sectoral inefficiency and rigidity losses and a change in the traditional ranking of industries in terms of their advantage in international trade.

Input-output tables are generally used to obtain estimates of the mechanical relative price changes that would be brought about by higher prices for energy.[29] It should be emphasized that computations of this type do not mean that such price changes will in fact take place in the

29. See, for Germany, Deutsches Institut für Wirtschaftsforschung, "Erdölkrise: Auswirkungen auf Preisniveau und Preisgefüge in der Bundesrepublik Deutschland," *Wochenbericht 3/74* (17 January 1974), pp. 16–17. For the United Kingdom, see *The Increased Cost of Energy: Implications for U.K. Industry,* a report by the National Economic Development Office, London, 1974. For France, see C. Gabet, G. Honoré, and F. Houssin, "Les répercussions mécaniques des hausses des prix énergétiques," *Economie et statistique,* 56 (May 1974), pp. 45–50. For Italy, unpublished estimates are provided in a dissertation by E. Clo and F. Paolucci presented in the Department of Economics, the University of Bologna, February 1975.

estimated magnitudes, or even according to the estimated ranking. In fact, they reflect only the mechanical transmission of higher import costs or production prices for energy. They do not consider the reaction of wages and salaries to higher prices. They are based on fixed technical coefficients in production, as well as on a fixed pattern of final demand. Flexibility in production and consumption would tend to moderate overall price increases but would also tend to blur the ranking of industries in terms of relative price competitiveness. On the other hand, transmission of price increases through wage-price spirals may accentuate the computed price increases and consequently change the ranking of industries even further.

Some additional considerations are in order in reference to the theoretical framework of the preceding section. First of all, it is clear that insofar as sectoral reallocation of production and demand brings about inefficiencies, these have already been counted in the aggregate within the macroeconomic terms-of-trade loss. It is only to the extent that there are rigidities in the process of sectoral reallocation that additional losses may occur. These are, in particular, losses caused by labor's inability to find at least temporary employment in other sectors.

The loss caused by inefficient production in energy-favored sections that expand is, as mentioned above, included in the aggregate terms-of-trade loss. It may also be overestimated, however, but for another reason. It is not to be excluded a priori that some of the sectors that will be developed as a consequence of higher energy prices, and in particular the energy-producing sectors, may hide infant-industry situations. In such cases, the initial real cost of domestic production is reduced with time and learning and correspondingly the overall inefficiency losses caused by reallocation of production that were measured earlier are themselves made smaller.

On the other hand, while true cases of infant industries will survive a return to cheaper energy prices, additional rigidity losses might appear if the price of oil were not stabilized at its new relative level. Indeed, the uncertainty surrounding the long-run price strategy of oil producers, while still causing terms-of-trade and rigidity losses, may be such as not to allow the investment that would bring about gains in infant industry. This may be particularly true for investment by the private sector, which is more sensitive to uncertainty than that in the public sector. Policy implications will be derived from these considerations later.

Macrodynamic Effects

The effects of the oil price increase on the distribution of income may be analyzed on the basis of a classic model of Ricardian tradition. In this model, land and labor are the original factors of production. The term *land,* of course, includes any fixed-supply, nonreproducible productive resource.

An open economy, by importing primary products, reduces the rent to be paid to domestic owners of land and, for given real wages, increases the rate of profit. In our case, the principal imported product under analysis is energy. An increase in its price would increase rents in domestic energy industries and in foreign energy-producing countries while lowering the overall profit rate in the domestic economy. Indeed, according to this theoretical framework, it is the cheapness of energy and of other primary products imported during the fifties and sixties that has allowed large profits in the economies of developed countries, even in the face of mounting real wages.

This simplified theoretical scheme indicates that, besides redistributing real income in the form of rents to oil-exporting countries and away from oil-importing countries, the rise in the price of oil will also affect the distribution of income among capital, labor, and domestic owners of fixed resources. For both reasons, while wages in the long run may be kept at their socially determined level, profit rates may fall in Western Europe and in other countries that are net importers of energy and primary resources.

A fall in the profit rate may reduce the rate of growth of the economy, unless the rate of saving rises substantially and aggregate demand is maintained at the full-employment level. While the rise in the rate of saving may be provided externally to Western Europe by members of OPEC, it is not certain that this would indeed occur if investment opportunities in Western Europe should yield a rate of profit lower than that yielded elsewhere.

Indeed, here is the root of a dilemma for OPEC. By having raised the price of oil they have very much increased their rents, but at the same time they have reduced the profitability of investing these rents in developed economies. Some of the gain to OPEC from higher oil prices may thus be offset by the induced fall in profits in developed countries and the consequent reduction in OPEC earnings on their invested surpluses. While empirical estimates cannot be made, it is almost certain, however, that the reduced earnings would offset only a small part of OPEC's gains in revenue from higher prices.

Policies for Reducing Long-Run Negative Effects

The analysis of the preceding sections may be used to provide some directions for long-run economic policy.

The first point concerns the need to avoid the possibility that the higher saving propensities of oil-exporting countries in relation to those of the oil-importing countries, and the redistribution of income toward the former, may provoke a fall in the long-run growth rate of the world. In itself, a larger rate of savings permits a higher growth rate, provided demand management is geared to full employment—that is, provided that investment opportunities for savings are attractive enough that savings will be channeled into the flow of expenditures. This requires also, besides the maintenance of full-employment income, good prospects of profitability for investment. Here again is the dilemma between larger rents, from which larger savings come, and lower profits on the investments that are supposed to attract those savings. To raise the profit rate, removal of certain taxes on business profits may be in order, as may tax discrimination in favor of foreign- (OPEC-) financed investment. In addition, the general level of profit in oil-importing economies should be maintained and increased by means of technological innovation, in an effort to offset the negative effect of higher energy costs on the profitability of energy-using processes.

This leads to a second set of policies connected with the protection of domestic energy production to offset uncertain and possibly falling prices for oil. In order to develop this production, and perhaps to obtain through technical progress the advantage of infant-industry cases, it is more expedient to use, as the theory indicates, direct subsidies to domestic energy-producing sectors, rather than to impose tariffs, variable levies, or quotas on imports of oil. In fact, the subsidy policy brings about only inefficiencies in production (which may disappear in cases of infant industries) while tariffs and quotas provoke welfare losses as well for the domestic users of energy.[30]

The introduction of tariffs or quotas cannot be justified by the need to break the oil cartel and lower the world price of oil, since a subsidy to domestic production of energy can always be designed in such a way as to produce the same reduction of imports as a quota or a tariff and thereby the same improvement in terms of trade. Thus, subsidies to domestic energy-producing sectors, besides having the advantage of recovering part of the terms-of-trade loss imposed by members of OPEC, would stabilize the long-run price to domestic producers, so as to allow reallocation of production and demand with a minimum of rigidity and uncertainty losses and a maximum of infant-industry gains. In any event, whether through production subsidies, tariffs, or quotas, a policy of reducing imports of oil would offer a significant advantage in that it would reduce the price of oil to developing countries that do not dispose of domestic energy resources. It would thus keep higher the profit rate for investments in those economies and with it their growth rate, provided sufficient savings were forthcoming to them from the developed countries and from members of OPEC.

The terms-of-trade loss may also be reduced by an indirect policy that does not require, as do those just outlined, a move to break the oil cartel and bring the price of oil down. As already explained, for a given world oil price, the larger the structural current-account deficit accepted by oil-importing countries and financed by OPEC members, the lower the terms-of-trade loss. This is still another reason (besides that concerning the need to keep a high rate of growth) for repealing taxes on foreign profits —to reward OPEC saving and thereby to reduce the terms-of-trade loss.

These policies, of course, are not limited in their application to the countries of Western Europe, but can be applied generally to most oil-importing countries. Clearly, the analysis should be pushed much further,

---

30. This does not mean that the subsidy is incompatible with taxes on domestic production, provided these are aimed at taxing away extra profits (rents) earned in expanding domestic production of energy. This seems to be the aim of parts of the recent tax proposals of President Ford. (Insofar as the objective of policy is solely to stimulate a given increase in domestic energy production, a subsidy will indeed accomplish that objective with less welfare loss than a tariff or quota. If, however, the objective of policy is to reduce imports by a given amount, a subsidy to domestic producers may be more or less costly than a tariff or quota depending, principally, on the size of energy supply elasticities in comparison to demand elasticities.—Eds.)

and the structural conditions of individual countries of Western Europe should be probed empirically in an effort to go beyond the level of generality of the considerations presented here.

Appendix: Derivation of Multipliers for Western Europe

Assuming that the increase in the import price of oil is the same for all Western European countries and that the marginal propensities to import oil (with respect to income) as well as the price elasticities of import demand for oil are not significantly different among the countries of Western Europe, it can be shown that the own multiplier for a group of countries (Western Europe) attributable to higher oil imports is

$$\frac{\Delta Y}{\Delta A} = \sum_i \sum_j k_{ij} w_i \qquad (i, j = \text{all countries in Western Europe})$$

where

$$w_i = Y_i / Y, \ Y = \sum_i Y_i$$

is Western European income,

$$\Delta A = \sum_i \Delta A_i$$

is the sum of the exogenous aggregate demand changes (represented by the net exogenous change in the trade account), and k_{ij} is the multiplier elasticity on country i's income due to a change in country j's aggregate demand,

$$k_{ij} = \frac{\Delta Y_i}{\Delta A_j} \bigg/ \frac{Y_i}{Y_j}.$$

Similarly, the multiplier effect of Western Europe on other countries (Japan and the United States for example) is obtained by aggregation and with the same simplifying assumptions about income propensities and price elasticities, so that

$$\frac{\Delta Y_q}{\Delta A} \bigg/ \frac{Y_q}{Y} = \sum_i k_{qi} + \sum_i \sum_{j \neq i} k_{qi} k_{ij},$$

where i and j are all countries in Western Europe, q is a country outside Western Europe (such as the United States or Japan), and, as before,

$$Y = \sum_i Y_i \quad \text{and} \quad \Delta A = \sum_i \Delta A_i.$$

Finally, the cross multipliers of other countries (Japan or the United States) upon the aggregate income of Western Europe are

$$\frac{\Delta Y}{\Delta A_q} \bigg/ \frac{Y}{Y_q} = \sum_i w_i(k_{iq} + \sum_{j \neq i} k_{ij}k_{jq}); \qquad w_i = Y_i/Y.$$

CHAPTER FOUR

Japan

TSUNEHIKO WATANABE
Osaka University

WELL BEFORE the oil crisis it had become evident in the early 1970s that the pattern of economic growth in Japan was changing. One of the forces making for change was public criticism of the cost of economic growth, expressed most notably in the concern over pollution and in pressure for more emphasis on social welfare expenditures. Another factor was the U.S. emergency economic measures of August 1971 and the adjustments in the international monetary system that they brought about, among the consequences of which was a substantial revaluation of the yen and an acceleration in the liberalization of Japan's import and foreign-exchange controls. These forces affected monetary and fiscal policies and these in turn had widespread economic effects. Private fixed investment, which had been the most important engine of growth in the 1960s, began to dampen, and Japan's unique price experience in the 1960s—a creeping inflation in consumer prices and stable wholesale prices—was replaced by a seriously rising trend in both wholesale and retail prices, with the former increasing even more rapidly than the latter.

The climate of uncertainty that followed the first revaluation of the yen in December 1971 caused government officials and businessmen alike to fear the possibilities of a prolonged and intensified recession. At the same time, the government came under growing public pressure to require in-

THE AUTHOR IS GRATEFUL for research assistance provided by Y. Ikeda and H. Suzuki of the Japan Economic Research Center and for comments received from other participants at the conference held in November 1974 at the Brookings Institution, where the paper was given—particularly for the comments of Hisao Kanamori. The final version of the chapter was completed in February 1975.

143

dustry to adopt antipollution measures, despite the cost-increasing consequences, and to expand welfare programs through public investment in
social overhead. In response to both forces, the new prime minister introduced expansionary fiscal and monetary measures, especially after the
middle of 1972. These were designed to restore the economy to the path
of 10 percent real economic growth that had characterized the 1960s. Thus
the official discount rate was reduced from 4.75 percent to 4.25 percent,
and a large supplementary government investment budget was authorized
by the Diet in October 1972. (Government investment was to rise by 29
percent for the fiscal year which had begun in April 1972 instead of the 20
percent increase originally provided for in the budget.) Additionally, exchange controls were relaxed, particularly as they applied to the outflow
of capital.

These expansionary measures seemed to succeed all too well. By the
first quarter of 1973 the economy had become overheated, with real
growth in gross national product (GNP) reaching an annual rate of 17
percent. Private fixed capital formation expanded rapidly, as did exports.

More of the same was in prospect for the fiscal year beginning 1 April
1973. The budget once again was expansionary and monetary policy relatively easy. (The official discount rate, while increased to 5 percent, was
significantly lower than in other advanced countries.) Demand had shifted
strongly toward consumer durables and luxury commodities, sales of
which were increasing at an annual rate of about 30 percent. This was reflected in such indexes as department store sales and in long waiting
periods for the purchase of new cars. Residential construction also
boomed, bringing about speculative increases in land prices and intensifying the supply shortages in the construction industry. And, despite the
discouraging impact of the further revaluation of the yen in the first
quarter of 1973 and the substantial cost of antipollution measures, almost
all the large corporations indicated that they were revising their investment
programs upward.

Inflation also accelerated. During 1973, the percentage increase in
the wholesale price index (WPI) and the consumer price index (CPI)
over corresponding months in the previous year showed a dramatic trend,
as follows:

	January	February	March	April	May	June	July	August	September
WPI	7.6	9.3	11.0	11.4	13.1	13.6	15.7	17.4	18.7
CPI	6.2	6.7	8.4	9.4	10.9	11.1	11.9	12.0	14.6

Thus the return to rapid real growth in GNP beginning in the second quarter of 1972 differed significantly from the pattern of growth during the 1960s, both in the existence of serious inflation and because the rise in wholesale prices, which was partly due to sharp increases in world commodity prices, had become stronger than that of consumer prices.

In part, the economic boom and the accompanying inflation from the end of 1972 onward stemmed from misjudgments in government policy. These included overstimulatory monetary and fiscal measures deriving from a mistaken forecast of the turning point from recession to rapid economic growth, an underestimate of the effect of sectoral shifts in demand resulting from government social investment programs, delay in controlling speculative activities brought on by excess liquidity, and delay in revaluing the yen.

Eventually the government began to adopt a restrictive stance in the management of demand. The Bank of Japan raised the official discount rate to 6 percent in July and to 7 percent in August. It also increased required reserve ratios and applied "window control"—that is, informal administrative restrictions on commercial bank lending. No significant changes, however, occurred in fiscal policy. Thus before the impact of the October oil embargo and the subsequent jump in oil prices, the forecast was that Japan's economic growth rate for the balance of 1973 and through fiscal 1974 would continue to be high and that the problem of controlling inflation would become more serious.

First Reactions to the Oil Embargo

Initially the Japanese government's reaction to the outbreak of the fourth Middle East war was comparatively sanguine. On October 6 the Ministry of International Trade and Industry (MITI) announced that there would be no immediate oil shortage since the national oil stockpile had been maintained at a level equal to 55 days' normal consumption. Pessimism, however, soon set in. By October 18, MITI had already decided to introduce nonmarket measures to reduce oil consumption. Prevailing opinion at this stage could be summarized thus:

• Sharp increases in oil prices would generate substantial deficits in the balance of payments, together with sizable devaluations of the yen; hence foreign reserves would shrink to less than $10 billion by March 1974 (from around $15 billion in October 1973);

• Production in major industries would be seriously constrained by oil shortages, thus bringing about a decline in GNP in the coming months and rising unemployment; and

• Even under these circumstances pressure on prices would become more intense, causing the rate of inflation in the Japanese economy to remain in two digits for the next few years.

Such pessimistic projections, together with concern over diplomatic developments in the Middle East, placed the government under strong political pressure to adopt measures that could maintain living standards. Furthermore, confidence in the ability of the market mechanism to control inflation, which already had been shaken by the events preceding the oil embargo, continued to deteriorate, even in the private business sector. If no action were taken, the combination of higher oil prices and restrictions on oil imports was expected to produce sharp increases in wholesale and consumer prices. These circumstances led MITI to propose direct government controls on the prices of major commodities and on the consumption of oil and electric power.[1] The emergency act authorizing these direct government controls was submitted at the end of November and passed by the Diet on 22 December 1973.

In addition to these direct controls on prices and consumption, the Ministry of Finance and the Bank of Japan actively adopted restrictive monetary and fiscal policies. Specifically, the budget for the fiscal year beginning April 1974 was significantly reduced from the size originally proposed in October 1973; the Bank of Japan in December 1973 raised the discount rate from 7 percent to 9 percent and applied window control to reduce the money supply; and exchange controls were tightened.

Measures to stimulate competition as a means of combating inflation constituted a third area for policy change. The Fair Trade Commission (FTC) became more active. Earlier, in the aftermath of the U.S. emergency economic measures, quasi-monopolistic behavior among the key industries such as steel and petrochemicals had been more or less condoned as necessary to adjust Japan's industrial structure to meet the new situation in the world economy caused by floating exchange rates. The advent of the oil crisis and the accelerated inflation that it brought about, however, built up support, especially among consumers, for more aggres-

---

1. The Ministry of Finance did not back these measures strongly, but MITI drew wide support for its proposals from recent experiences with panic buying and consequent sharp price rises for toilet paper, detergents, and other consumer goods.

sive action by the FTC. The oil refineries were the first target; in February 1974 the FTC undertook to prosecute certain major oil-refining corporations for collusion in fixing prices. In the following months it made several investigations of monopolistic behavior elsewhere and issued a number of administrative warnings.

Finally, the Diet began to inquire actively into possible administrative mismanagement in the government and malfeasance on the part of large business enterprises, although for the most part these inquiries proved to be politically motivated and not particulary effective.

The three principal approaches—price and allocation controls, restrictive fiscal and monetary measures, and the promotion of competition—were not initiated as part of a consistent plan, largely because of the initial confusion surrounding the current and prospective availability of oil. Since Japan is almost wholly dependent on imports for its energy supply, particularly in the case of oil, the future supply of imports is necessarily a crucial determinant of industrial production.[2] From the beginning, uncertainty and conflicting information existed about how much oil was available, how much was being and would be imported, and the extent to which prices would rise—questions that at the time proved to be equally difficult to answer in other industrial countries. For example, in the middle of December 1973 the amount of oil that would be available for import during the first quarter of 1974 was variously projected to be down by 5 percent to 30 percent from originally anticipated levels. In fact the amount of oil actually imported in November and December 1973 was not known with any certainty until the latter part of January 1974. This confusion about the facts was frequently attributed to government mismanagement, but the important consequence was that it complicated the government's task of developing effective and consistent policy measures.

Initially, also, the oil embargo had an exaggerated impact on prices of other commodities. As described in the previous section, the wholesale and consumer price indexes were rising rapidly before the oil crisis. The embargo added impetus to the inflation, however, by creating fears of a general shortage of supplies—fears which reached near panic proportions. In the middle of November the supermarkets witnessed a run on a variety of consumer goods, despite the fact that supplies of these commodities were unrelated to oil, and this panic buying caused prices to jump in some

2. The supply of oil has become a crucial input even for agriculture, as a result of the growing mechanization of production.

instances by roughly 50 percent. MITI eventually brought the situation under control by emergency allocations of supplies, but the episode left a residue of higher prices.

After the disappearance around the end of January 1974 of panic behavior among consumers and confusion about how much oil would actually be available, the reaction to the oil crisis moved into a second stage. This stage was characterized by two governmental adjustments. The first consisted of actions to capture or eliminate windfall profits resulting from the pricing policies and speculative behavior of a number of large business enterprises. These had been strongly criticized in the Diet for their lack of social responsibility. The second adjustment concerned action to permit the price system to reflect the fundamental change that had occurred in primary commodity prices. Instead of enjoying low or declining prices for primary commodities, as had been the case during the 1960s, Japan, it was believed, would now have to face permanently high relative prices for energy resources, raw materials, and food.

The recapture of windfall profits was sought principally by the imposition of a 10 percent surcharge on corporate profits, beginning in March 1974.[3] Revenues to be realized in fiscal 1974 from this surcharge were expected to be 170 billion yen.

The second adjustment was seen as having two stages: in the first, direct government control of prices (the price freeze) was to be eliminated by September 1974, while restrictive fiscal and monetary measures were still maintained; in the second, a "soft landing" upon a desirable path of economic growth was to be achieved. This pattern might consist of 5–7 percent real growth in GNP, a lower, single-digit rate of inflation, and a reasonable current-account surplus in the balance of payments. Although such goals were not explicitly formulated, it was expected that something along these general lines might be achieved by the end of 1975.

Elimination of direct government control of prices actually began at the end of March 1974 with the authorization of an average 62 percent rise in prices of oil products. Subsequently price increases were authorized for dairy products (40 percent), electric light and power (57 percent), ben-

---

3. The surcharge will be in effect for fiscal years 1974 and 1975. It is to apply to profits exceeding 500 million yen or to profits in excess of 20 percent of paid-in capital. Small firms are exempted. In all, some 1,600 corporations are expected to pay the tax. In addition, government financial institutions, through their loan policies, imposed pressure on several corporations.

zene, toluene, xylene (35 percent each), liquefied petroleum gas (39 percent), steel products (17 percent), and private railway fares (26 percent). In addition, an average wage increase of 33 percent was agreed upon in the wage negotiation in the spring of 1974, and the price of rice received by the farmer was increased by 40 percent.

As price controls were removed, the gradual return to a desirable pattern of economic growth would depend principally on fiscal and monetary measures. Once these adjustments had been made, the impact of the oil shortage and higher oil prices on Japan's economy could be said to have disappeared.

The following sections of this chapter seek to quantify the impact of higher oil prices on Japan's inflation and economic growth and to assess the possibility for achieving a more desirable pattern of economic growth within the next few years.

The Effect of Rising Oil Prices on the Japanese Economy: October 1973 to March 1975

As pointed out in the previous section the Japanese economy, even without the oil crisis, was experiencing a substantial inflation in 1973 and was heading for a period in which strong fiscal and monetary measures would be required to bring that inflation under control. Some tightening of money had been undertaken in the summer of 1973. Sole reliance on monetary policy to deal with inflation, however, had been criticized by many nongovernment economists, who urged that fiscal policy also be made more restrictive during the coming 1974 fiscal year (April 1974 through March 1975). Virtually every expert was in agreement that the inflation in wholesale prices could not be brought down below the two-digit level until some time in the middle of 1974, and consumer prices only by some time in 1975.

Needless to say, given the inflationary environment described above, it was certainly desirable to maintain the combination of a smaller budget with tighter monetary measures, at least from the point of view of national economic management. Japan's political situation at that time, however, especially with the election of the House of Councilors scheduled for July 1974, made it difficult for this line of policy to be carried out, particularly if it meant that the rate of economic growth (in the terms of the real

GNP) would have to fall below 5 percent per annum. Indeed, if the country's experience in the general election and the election of the House of Councilors over the past twenty years were taken as a guide, the party in power, the Liberal Democratic party, was likely to seek victory in the election by means of a sizable reduction in income taxes and an expansion of selected fiscal outlays. Even in the face of the prevailing and predicted strong inflationary tendency, it was almost impossible, for domestic political reasons, to expect any significant change in this strategy. That is to say, there was little hope that economically desirable adjustments of policy, namely the introduction of restrictive fiscal measures, would be adopted with any consistency during the pre-election period stretching from October 1973 to June 1974. In fact, as early as September 1973 the prime minister had already held out the hope of a reduction of two trillion yen in the income tax without explicitly discussing commensurate reductions in fiscal outlays. (Fulfillment of this political commitment in April 1974, even though accompanied by an increase in the corporate profits tax, diluted the effect of the government's post–October 1973 restrictionist measures.)

The oil crisis, therefore, impinged upon a situation in which economic policy, especially fiscal policy, could be projected as insufficiently restrictive to deal with the existing inflationary pressures. The sharp jump in oil prices beginning in the fall of 1973, however, led the government to adopt further restrictions in monetary policy and to undertake a series of measures designed to tighten fiscal policy. Two sets of forces—higher oil prices and anti-inflationary policy measures—therefore began to affect the Japanese economy in late 1973 and throughout 1974.

An Overall Appraisal

In order to disentangle the various factors at work we have used our econometric model to simulate the behavior of the Japanese economy from April 1973 through March 1975 under three different sets of assumptions.

1. A *no-change* model, which assumes a continuation of the fiscal and monetary policies which appeared to be emerging before the oil crisis and no rise in oil prices from mid-1973 levels;

2. A *policy-change-only* model, which takes into account the effect of

Table 4-1. *Major Assumptions Used in Three Economic Models*

Variable and model	1973:4	1974:1	1974:2	1974:3	1974:4	1975:1
Volume of world trade (annual percentage rate of change)						
Model 1	12.0	11.4	9.0	9.0	9.0	9.0
Model 2	10.8	3.0	6.0	6.0	6.0	6.0
Model 3	10.8	3.0	6.0	6.0	6.0	6.0
World price index of manufactured commodities (annual percentage rate of change)						
Model 1	9.0	6.5	6.0	6.0	6.0	6.0
Model 2	9.0	6.5	6.0	6.0	5.8	5.8
Model 3	15.2	20.0	20.0	20.0	15.0	10.0
Implicit price deflator for imports of goods and services (annual percentage rate of change)						
Model 1	15.0	15.0	15.0	15.0	15.0	10.0
Model 2	15.0	15.0	15.0	15.0	15.0	10.0
Model 3	26.6	49.4	59.8	47.0	40.0	15.0
Official discount rate, Bank of Japan (percent)						
Model 1	7.0	7.0	6.5	6.5	6.5	6.5
Model 2	7.2	9.0	9.0	9.0	9.0	8.0
Model 3	7.2	9.0	9.0	9.0	9.0	8.0
Increase in money supply (annual percentage rate of change)						
Model 1	20.0	19.0	18.0	18.5	19.6	18.6
Model 2	16.8	15.4	15.7	13.0	15.0	16.0
Model 3	16.8	15.4	15.7	13.0	15.0	16.0
Increase in real value of government consumption (annual percentage rate of change)						
Model 1	4.3	11.1	4.3	4.3	4.3	4.3
Model 2	5.5	2.6	1.7	1.5	2.3	3.8
Model 3	5.5	2.6	1.7	1.5	2.3	3.8
Exchange rate (yen per dollar)						
Model 1	270	265	265	265	265	265
Model 2	272	290	280	300	280	280
Model 3	272	290	280	300	280	280

the policy changes introduced after the oil crisis but assumes no rise in oil prices.

3. A *policy-change-plus oil-crisis* model, which incorporates not only policy changes but higher oil prices.

Table 4-1 summarizes the specific assumptions used in each model. Table 4-2 presents the results in terms of percentage changes in the major economic variables between fiscal 1972 and fiscal 1974. In this table a comparison of Models 2 and 3 provides estimates of the impact on Japan of the oil price increases alone, given the policy measures that were taken.

Table 4-2. *Changes in Major Economic Variables, Fiscal 1972 to 1974*

Variable	Model 1: No change (1)	Model 2: Policy changes only (2)	Model 3: Policy changes plus oil crisis (3)	Differences Higher oil prices alone (3−2)	All changes (3−1)
			Percent		
Real consumption	18.5	16.0	10.9	−5.1	−7.6
Real fixed investment	21.4	8.4	4.7	−3.7	−16.7
Real housing investment	3.2	−4.4	−6.3	−1.9	−9.5
Real exports	23.8	26.9	31.4	4.5	7.6
Real imports	44.4	39.8	39.5	−0.3	−4.9
Real GNP	15.4	11.4	6.0	−5.4	−9.4
Current GNP	47.8	43.5	52.3	8.8	4.5
Wage per employee	45.6	43.5	51.7	8.2	6.1
Export price	28.3	31.8	47.0	15.2	18.7
Consumer price index	22.3	22.0	42.3	20.3	20.0
Wholesale price index	32.3	31.6	55.3	23.7	23.0
Labor productivity	13.1	9.3	4.2	−5.1	−8.9
			Billions of dollars		
Change in the current balance	−1.1	4.8	−8.3	−13.1	−7.2

A comparison of Models 1 and 3 shows our estimates of the combined effects of both the policy changes and the higher oil prices.

The direct and indirect effects of the spurt in oil prices on the rate of inflation have been quite serious. What would, in any event, have been a significant inflation, even with more restrictive policies, was made still worse. Over the two-year period, wage increases of 46 percent accelerated to 52 percent. A 32 percent increase in wholesale prices was raised to 55 percent, and a 22 percent consumer price rise grew to 42 percent. The major feature that distinguishes the assumptions used in Model 3 from those in Model 2 is the accelerated rise in the import price index following the jump in oil prices which occurred in October 1973 and again in January 1974. While the oil price rise did contribute, directly and indirectly, to the sharp increase in Japanese import prices in late 1973 and throughout 1974, it was not the only cause of that rise. Hence, the economic effects that we have attributed to oil prices are, in actuality, also attributable to the more generalized spurt in world commodity prices, on top of the rapid increases already occurring in 1973.

The real growth in GNP, which might have reached 15 percent over the two-year period, was substantially slowed. Restrictive monetary and fiscal policies alone would have reduced growth to 11 percent (over two years). The drain of purchasing power caused by the higher oil prices slowed growth still further. Indeed, after a growth of 10 percent in 1973, GNP actually fell by more than 3 percent in 1974; for the two years combined, therefore, the total rise in GNP was only 6 percent, an insignificant rise by postwar Japanese standards.

The policy measures, taken alone, had their major effect on business investment and housing, as the tight money measures bit into the spending of these sectors. Consumption was for the most part untouched by the policy measures but was severely hit by the rise in oil prices, as consumers, having paid the higher prices for oil products, were left with less real income for purchasing other goods and services. The prospective decline in foreign reserves, which might have been reversed with the policy measures, was made substantially worse by the oil deficit. The current balance in Model 3 declines by $8.3 billion over the two-year period, compared to a prospective rise of $4.8 billion if the policy measures had been taken but no oil price rise had occurred. The $13 billion decrease in the foreign balance, caused by higher oil prices, is about equal to the increase in the "oil deficit." Over this period the added cost of oil imports, less added exports to countries of OPEC, will amount to perhaps $14 billion.

The Quarterly Pattern

We have also simulated the effects of policy changes and higher oil prices on the quarterly patterns of economic activity, from the fourth quarter of calendar 1973 through the third quarter of 1974. Table 4-3 summarizes the results for Model 1 (no change) and Model 3 (policy changes plus higher oil prices) during that period.

The quarterly simulations show how the Japanese economy was adapting, or failing to adapt, to its economic problems during the period. The most severe loss of growth in output, in relation to the base case of Model 1, occurred in the first half of 1974, when declines in output were experienced. By the third quarter, output had stabilized but was still below what could have been expected under a continuation of earlier conditions. Within the total GNP, however, the impact on investment became increasingly severe during the year, while the impact on consumption was moderating. The higher oil prices caused the foreign balance on current account

Table 4-3. *Simulation of Quarterly Impact of Policy Changes and Higher Oil Prices on the Japanese Economy*

Variable and model	1973:4	1974:1	1974:2	1974:3
	Percent change over previous year			
GNP (*constant prices*)				
Model 1	9.3	3.6	6.6	5.7
Model 3	5.7	−4.2	−2.8	0.0
Private consumption (*constant prices*)				
Model 1	9.8	6.0	8.4	10.9
Model 3	8.2	0.3	0.4	3.7
Private fixed investment (*constant prices*)				
Model 1	21.0	0.6	4.6	1.1
Model 3	21.0	−0.4	−6.5	−13.2
Exports (*constant prices*)				
Model 1	−2.5	6.2	20.9	13.1
Model 3	1.9	6.8	22.7	24.2
Imports (*constant prices*)				
Model 1	28.9	28.4	18.7	17.3
Model 3	28.0	24.2	13.5	14.4
Consumer price index				
Model 1	16.0	18.8	12.3	5.9
Model 3	16.4	24.4	23.8	23.1
Wholesale price index				
Model 1	18.9	20.0	13.6	14.2
Model 3	24.0	35.5	35.5	32.1
Earnings per employee				
Model 1	25.2	15.6	22.4	20.2
Model 3	25.3	17.7	25.5	24.6
	Millions of dollars			
Foreign balance on current account				
Model 1	−130	−1,700	−130	−240
Model 3	−330	−3,350	−2,460	−1,700

to deteriorate substantially. But the combination of policy measures and lower GNP began to stimulate exports and reduce imports. As the year wore on, therefore, the deterioration in the current-account balance was moderated, and by the third quarter of 1974 a significant part of the oil deficit had been offset by improvements elsewhere in the current account.

While the effect of the oil crisis on real output and the balance of payments gradually lessened, the impact on prices became worse during the year. With a lag, price increases for imported commodities spread through the economy. Instead of a gradual improvement in consumer price inflation to a 6 percent annual rate in the third quarter, there was a rise of 23 percent in prices. A somewhat more modest, but still significant, improvement in wholesale price behavior was converted to a 32 percent rate of inflation. The acceleration of wage increases was much less, but this is partly explained by the fact that wage increases for the year were set by negotiations in the spring of 1974, before the full effects of consumer price inflation had been felt.

In summary, the impact of the oil crisis and other developments to which it led were severe on total output and the balance of payments. But by the third quarter, these effects—while still large—were already lessening. The sharp and continuing decline in fixed investment, however, could pose a major threat to future growth. And the inflationary impact, through cost increases, was becoming worse as the year wore on, providing a major problem for the design of future economic policy.

The Impact of Oil Supply Reductions: A Hypothetical Case

As noted earlier, considerable pessimism existed in December 1973 as to how much oil Japan would receive under the embargo. The most pessimistic prediction suggested an enforced reduction in supply of 20 percent for the period October 1973 to March 1974. For the fiscal year 1974 (April 1974 through March 1975) a reduction of about 4 percent from expected requirements was forecast. While these shortfalls in oil imports and supplies did not in fact materialize, it is useful, as a means of tracing the importance of oil in Japan's economy, to quantify their possible economic impact.

Toward this end, we have used our econometric model to simulate the behavior of the Japanese economy over the period 1973–75 in circumstances in which the supply of oil would have been reduced. In other words, how much worse off would the economy have been if, in addition to higher oil prices, the supply of oil had been forcibly reduced?

For purposes of these simulations, variations in the volume of oil imports in fiscal years 1973, 1974, and 1975 are assumed to be as follows (in millions of kiloliters):

Oil-supply assumption	*1973*	*1974*	*Percent change*	*1975*	*Percent change*
Originally scheduled	303	330	9.0	...	...
Actual	289	306	6.0	339	11.0
Pessimistic	272	260	−4.0	260	0.0
Moderate	276	266	−3.5	280	5.3

"Originally scheduled" refers to planned import requirements before the oil crisis; "actual" means the oil imports realized in 1973 and projected for 1974 and 1975, assuming no constraints; "pessimistic" refers to the lowest estimates made during December 1973 in the face of the oil crisis; and "moderate" means simply a somewhat less pessimistic forecast. Columns 3 and 5 give the percentage by which the volume of imports in 1974 and 1975 respectively represent an increase or decrease from the preceding year's volume.

In order to deal explicitly with these differing oil-supply assumptions, our model has been redesigned in the following ways: variations in industrial production are accounted for directly by variations in oil supply; the gap between the newly constrained supply of industrial output and the demand for industrial output that the existing economic structure would have generated is adjusted directly by changes in all endogenous economic variables; and finally, speculative inventory accumulation resulting from the prevailing negative rate of interest is explicitly taken into account.[4]

Assumptions with respect to exogenous variables and other structural equations are the same for the three supply cases listed above.[5]

Let us review yearly results first. Compared with the situation in which there are no constraints on oil imports (Case A), the most pessimistic

4. Technically, our model has been redesigned as follows: to estimate a supply function in manufacturing production, $O^s = f$ (import of oil); changes in inventory investment are computed from $J^* = J_0 + \Sigma \Delta J$, where J_0 is the initial estimate of inventory investment based upon our previous specification, $\Delta J = \lambda(O^s - O^d)$, where O^d is determined from the original function such that $O^d = f(C, I, E, \ldots)$; and λ is a positive value between 0 and 1. Any given ΔJ is equal to zero when $O^s = O^d$.

Specifically the following equations are newly estimated and added to the original macro model.

$$\log O^s = -3.1532 + 0.41720 \log (M_{\text{oil}}) + 0.29814 \log (M_{\text{oil}}) - 1$$
$$J_0 = J$$
$$J^* = J_0 + \Delta\lambda(O^s - O^d)$$

where M_{oil} = the amount of oil import, J = changes in inventories in the original model, J^* = the newly adjusted amount of inventory change.

5. The base forecast, or the amount of the imports originally scheduled before the oil crisis and before policy changes, was discussed in the previous section.

supply forecast (Case B) would have reduced real growth of GNP by an average of 2.5 percent a year over the period 1973–75 (Japan's fiscal years). Even a moderate shortfall in the availability of oil (Case C) would have brought about a reduction in real growth of GNP by an average of 1.2 percent a year. The results of these calculations of the growth of real GNP during the three fiscal years in question are as follows (in percentages):

Oil-supply assumption	1973	1974	1975	Average, 1973–75
Case A	5.5	0.5	2.1	2.7
Case B	0.7	−3.0	2.8	0.2
Case C	1.3	−2.3	5.4	1.5

Marked differences exist in the impact of reduced oil supplies on the major components of the GNP, as shown in Table 4-4. Investment would have been the most drastically changed. Indeed, these calculations suggest that enforced reductions in oil supply, whether extreme or moderate, would soon have brought about negative growth in fixed investment—the most important force for economic growth in postwar Japan. Inventories would have dropped sharply in the first year of reduced oil supplies and been rebuilt rapidly in the third year, contributing to a substantially different cyclical pattern from that which in fact occurred. The current balance would also have been substantially different in the restricted supply cases, moving strongly into surplus, almost entirely because of reductions in the volume of oil imports. On the other hand, as might be expected, prices on a yearly basis would not have been affected, since by assumption the import price of oil would be the same in all three supply cases. Wage increases, however, would have been somewhat smaller in a situation of reduced oil supplies because of the lower rates of growth in GNP associated with supply restrictions.

Some differences in emphasis and timing emerge from simulations made on a quarterly basis. Three quarterly periods have been used: October–December 1973, January–March 1974, and April–June 1974. The calculations indicate that reduced oil supplies would have depressed the growth of private consumption, private investment, inventories, and imports. As can be seen from the simulations in Table 4-5, reductions in oil supply would first have brought about reductions in consumption, imports, and inventories and then would have weighed heavily on private fixed

Table 4-4. *Impact of Reduced Oil Supplies on Major Components of GNP during Three Fiscal Years*

Variable and oil-supply assumption	1973	1974	1975
Growth of fixed inventory, based on 1965 prices (percent)			
Case A	14.4	−8.5	5.1
Case B	10.9	−42.9	−28.5
Case C	10.9	−41.6	−25.0
Inventory (trillions of yen)			
Case A	3.10	3.90	2.21
Case B	0.39	3.05	3.76
Case C	0.74	3.85	5.86
Current balance (billions of dollars)			
Case A	−4.2	−4.1	−0.0
Case B	−1.6	1.2	6.6
Case C	−1.9	0.4	4.0
Annual wage increases (percent)			
Case A	20.7	25.7	21.9
Case B	20.4	21.4	14.5
Case C	20.4	21.8	14.9

investment. The decline in growth rates of real GNP would have been immediate and substantial.

In addition, foreign reserves would have increased by $3 billion over this nine-month period, but neither prices nor the wage increases negotiated in the spring of 1974 would have been affected.

In sum, an embargo that substantially reduced the availability of oil to Japan would have curtailed private investment drastically and would have had widespread effects on the economy of Japan. If these effects had been added to the consequences of higher oil prices, discussed in the previous section, the Japanese economy might well have been forced into a serious depression.

Impact on the Structure of Prices

In addition to their impact on the general price level, higher oil costs could be expected to change the relative price structure. Not only would energy itself become more expensive in comparison to other goods and

Table 4-5. *Quarterly Simulations of Annual Changes in Real Rates of Growth*

Variable and oil-supply assumption	1973:4	1974:1	1974:2
		Percent	
Private consumption			
Case A	8.2	0.3	0.4
Case B	5.8	−2.3	−0.5
Case C	5.8	−1.9	−0.4
Private fixed investment			
Case A	21.0	−0.4	−6.5
Case B	21.0	−13.1	−37.2
Case C	21.0	−13.1	−31.4
Imports			
Case A	28.0	24.2	13.5
Case B	23.2	17.7	8.5
Case C	23.2	18.8	9.3
Real GNP			
Case A	5.7	−4.2	−2.8
Case B	−2.3	−14.3	−10.2
Case C	−2.3	−12.2	−9.0
		Trillions of yen	
Changes in inventories			
Case A	0.80	0.53	1.27
Case B	−0.65	−0.74	0.95
Case C	−0.65	−0.39	0.98

services, but those commodities whose production required much energy or petrochemical feedstocks would show a greater rise in price than would other goods using less of these materials. To see how large these relative price effects might be we have traced through the consequences of a rise in oil prices from $3 to $9 per barrel using an input-output model. The specific assumptions and methodology of this exercise are shown in the appendix to this chapter, and the principal results, for selected industries, are given in Table 4-6.

Each percentage in Table 4-6 represents the increase in the average supply price of the particular industry consequent upon a rise in oil prices from $3 to $9 per barrel,[6] with the price rise beginning in the first quarter

6. Both the $3 and $9 prices per barrel represent prices for Saudi Arabian oil, f.o.b. Persian Gulf.

Table 4-6. *Impact of Increase in the Price of Oil from $3 to $9 per Barrel on Selected Wholesale Prices*[a]

Sector	1974	1975	1976
	Percent differences in prices		
All commodities	6.6	7.4	7.6
Petroleum and coal products	78.9	83.2	83.6
Iron and steel	7.9	9.4	9.6
Chemicals	4.7	5.7	5.7
Nonmetallic minerals	3.6	4.2	4.2
Pulp and paper products	3.0	3.6	3.6
Metal products	2.8	3.4	3.4
General machinery	2.2	2.6	2.6
Transport equipment	2.2	2.4	2.4
Lumber and wood products	1.2	1.4	1.4
Foodstuffs	1.0	1.2	1.2

a. See the appendix to this chapter for a discussion of the methods used to produce these results.

of 1974 and being passed through into product prices on an absolute basis, without further markup. In this simulation no allowance is included for further indirect effects through the wage-price spiral. The estimates assume that the rise in Persian Gulf prices would be fully reflected in delivered oil prices by the third quarter of 1974 and in manufactured gas prices by the fourth quarter.

As Table 4-6 indicates, this tripling of oil prices, when passed through into product prices, would over the period 1974–76 raise the wholesale price index by almost 8 percent over the level it would otherwise have reached. Petroleum and coal product prices, on the average, rise by approximately 85 percent. Iron and steel prices, because of the increase in coking coal prices associated with higher petroleum prices, increase by some 10 percent and chemical prices by 6 percent. Other oil-induced price changes would be more modest. Within these broad industry aggregates, of course, the relative prices of individual commodities would be more severely affected. Nevertheless, outside of petroleum products themselves, and iron and steel and chemicals, shifts in the relative price structure do not appear to be very large.

In calculating the effects of the oil price increase, each of the two projections ($3 and $9 a barrel) was estimated in the context of a fuller set of assumptions about wages and about prices of imports, rice, and trans-

Table 4-7. *Simulation of Price Increases for Selected Groups of Industries, 1972–76*[a]

Sector	1972–73	1973–76	1972–76
		Percent change	
Petroleum and coal products	7.8	183.1	205.1
Nonferrous metals	30.9	95.1	155.3
Ferrous ores and scrap	25.8	85.3	133.1
Lumber	44.2	30.7	88.5
Metal products	12.7	67.1	88.3
Chemicals	12.8	62.0	82.7
All commodities	15.8	57.4	82.2
Pulp and paper	21.0	47.5	78.5
Primary iron and steel	15.3	52.1	75.3
Textiles	36.5	27.5	74.1
General machinery	10.3	49.1	64.4
Foodstuffs	10.5	47.0	62.5
Transport equipment	2.1	44.5	47.5
Electrical machinery	1.8	44.2	46.7

a. See the appendix to this chapter for an outline of methodology and assumptions. Industry groups are arranged in descending order of the magnitude of their simulated price increases during the entire period 1972–76, as shown in Column 3.

portation, taking into account the large import price and wage increases which occurred in 1973 and 1974 and projections of further changes in 1975 (as outlined in the appendix to this chapter). Table 4-7 shows an input-output simulation of the percentage increases in prices from 1972 to 1976 associated with an oil price of $9 a barrel and the recent (and projected) rise in prices of rice, transportation, and imports and in wage rates. Apart from petroleum and coal products, relative price changes are, not surprisingly, dominated by the degree of fabrication involved in the product in question. This stems from the fact that wages, which are an increasingly important component of costs, rise much less than the prices of raw materials. As a consequence, at one end of the scale ferrous and nonferrous primary metals rise by 133 and 155 percent respectively, while at the other end, such highly fabricated products as machinery and transport equipment rise by very much smaller amounts. Price increases for intermediate products—chemicals, textiles, pulp and paper, and fabricated metal products—lie between the two extremes.

Table 4-8. *Projections of Oil Imports and Associated Changes in Real GNP*

Year and quarter	Oil imports		Percent change in real GNP	Price of oil in U.S. dollars per barrel
	Millions of kiloliters[a]	Percent change		
1974:1	69.9	...	−4.2	8.9
2	72.4	3.3	−2.8	11.2
3	75.1	−0.4	0.0	11.7
4	79.3	8.1	−0.1	12.1
1975:1	76.5	9.4	5.3	12.8
2	72.7	0.4	1.8	12.5
3	76.3	1.6	1.3	13.2
4	82.0	3.5	2.5	13.8
1976:1	79.6	4.0	2.9	14.6
2	76.0	4.5	3.2	14.8
3	80.7	5.9	3.9	15.0
4	88.0	7.3	4.4	15.7
1977:1	86.2	8.4	5.3	16.5
2	85.3	12.2	7.5	16.5
3	92.2	14.3	8.9	16.6
4	101.3	15.1	9.5	17.4

a. Percentage rates of change for the volume of oil imports and for real GNP represent annual rates of change from the corresponding quarters of the previous year.

Balance of Trade between Japan and Members of OPEC

A further aspect of the analysis concerns the effects of higher oil prices on Japan's trade balances with members of OPEC. These changes are projected for the period 1974 through 1977.

In Table 4-8, Japan's oil imports are estimated on the basis of our economic forecast for Japan for this period and on the assumption that the real price of oil will remain at the average level for 1974 (with adjustments of the dollar price in subsequent years corresponding to the assumed rate of world inflation).[7]

7. By using quarterly data from 1965, the following oil import function from OPEC members has been estimated and introduced into our macro-model in order to estimate required amounts of imported oil for 1975–77 simultaneously with other estimates of aggregate variables:

$$\log M = 6.513 + 1.518 \log O - 0.146 \log (P_M/P_j)$$
$$(33.4) \qquad (-2.5)$$

where M = oil imported from OPEC members in 1965 constant prices, O = produc-

Japan's exports to members of OPEC increased rapidly during the first three quarters of 1974. Compared with the corresponding quarters of the preceding year these increases amounted to 54.3 percent in the first quarter, 90.7 percent in the second quarter, and 100 percent in the third quarter. Our projection of exports to members of OPEC for 1975–77 includes specific assumptions about changes in Japan's terms of trade with these countries, in its competitive position compared with other industrial countries exporting to OPEC markets, and in the level of world trade. This equation for exports to members of OPEC gives the following results (in billions of United States dollars at constant 1974 prices):[8]

1973	2.96
1974	4.63
1975	5.53
1976	6.20
1977	7.16

In terms of current prices the trade balance, on a customs-clearance basis (exports f.o.b., imports c.i.f.), would be as follows (in billions of United States dollars):

	Imports	Exports	Balance
1973	6.42	2.54	−3.88
1974	18.95	4.63	−14.29
1975	23.02	6.43	−16.59
1976	27.34	7.92	−19.42
1977	34.32	9.95	−24.37

Thus in current dollars Japan's trade deficit with members of OPEC would increase by approximately $10 billion in 1974 as a result of higher oil prices and by an additional $10 billion over the succeeding three years. In constant 1974 dollars the trade deficit would increase from $14 billion in 1974 to $17 billion in 1977, despite policies to reduce demand and promote exports.

---

tion index, P_M = oil price, P_j = deflator of inventories. The figures within the brackets are the corresponding t-values. The variables O and P_j are endogenously determined within the macro-model system. (Because of differences in methodology, the growth of Japanese exports to members of OPEC estimated by Professor Watanabe is significantly smaller after 1974 than the growth estimated in Chapter 1 [see Table 1-3].—EDS.)

8. The figure for 1973 represents actual exports. These estimates are simultaneously determined with other economic variables in the model by introducing the export equation for OPEC members into the simultaneous equation system.

Conclusions

As we have seen, the oil crisis impinged on a Japanese economy that was already suffering from substantial inflation. Consumer prices were increasing rapidly and, contrary to the pattern of rapid economic growth in the postwar period, industrial wholesale prices were also rising sharply. The additional spurt in prices and wages set off by the oil crisis not only had its direct effects on the economy but also—after an initial period of direct controls—led to the adoption of more restrictive monetary and fiscal policies.

A significant cut in the physical supply of oil together with higher oil prices would have had major adverse consequences for the Japanese economy. Fortunately, such shortages did not occur. But the drain of purchasing power that accompanied higher oil import prices, the restrictive monetary and fiscal policies subsequently introduced in order to fight inflation, and the general uncertainty engendered by the embargo did sharply reduce the growth of output, especially during the middle of 1974.

Our estimates imply that by March 1975 the rates of price increase in the WPI and CPI would recede to those prevailing just before the October 1973 embargo and that foreign reserves would be restored to the October 1973 level. At some risk of overstatement these conclusions suggest that the Japanese economy could be expected to have absorbed, and to have adjusted to, the effect of higher oil prices on cost-push inflation over a period of about fifteen months. Sustaining restrictive monetary and fiscal policies for that length of time would presumably be sufficient to overcome the effects of the oil crisis alone. But since a substantial inflation was under way *before* the oil crisis began, a return to the pre-embargo situation would not suffice.

In order to insure a soft landing of the Japanese economy—that is, a reduction of inflation below the double-digit level—continuation of restrictive monetary and fiscal policies may be necessary for the next several years. Yet, as Table 4-9 indicates, even with a continuation of those policies, double-digit inflation in consumer prices is likely to persist through 1977, and the inflation in wholesale and export prices may not drop significantly below double-digit levels until that year. Meanwhile, given the monetary and fiscal policies necessary to insure a soft landing, the growth in real GNP may continue at a subnormal level, at least through 1976.

Under the present political situation in Japan, characterized by the

Table 4-9. *Projected Consequences of Current Monetary and Fiscal Policies*

Percent change from preceding year; fiscal years

Variable	1974	1975	1976	1977
Real GNP	−0.5	2.1	4.2	7.9
Consumer price index	23.1	16.6	13.6	12.3
Wholesale price index	26.7	11.1	9.7	9.1
Wage per employee	25.7	21.9	15.8	17.8
Export price index	25.9	22.0	9.5	8.7

narrow seven-seat majority of the Liberal Democratic party in the upper house of Parliament, it is highly unlikely that an explicit form of income policy or of wage-price controls could be introduced successfully. Barring a significant decline in energy prices, only restrictive monetary and fiscal policies are available to cope with the inflationary problem. And such policies are likely to produce rising unemployment, perhaps to a point where more than one million persons are unemployed.

How would lower oil prices affect this outlook? Suppose, for the sake of illustration, that oil prices (in constant 1974 dollars) gradually fell from $10 in 1974 to $6 in 1977. Such a decrease in oil prices would reduce the assumed rate of growth in the overall import price index used in making the projections of Table 4-8 by 2.0, 3.3, and 4 percent in 1975, 1976, and 1977 respectively. An estimate of the impact of these changes on inflation, using an approximation based on a truncated version of our econometric model,[9] would be as follows, in percentage changes for fiscal 1975 and 1976:

	Consumer price index		Wholesale price index	
	1975	1976	1975	1976
$10 price continued	17	14	11	10
Price falls to $6	14	12	10	9

The effect is a one-year shift in rates of price increase—that is, the deceleration of price inflation is achieved one year earlier. As a consequence restrictive monetary and fiscal policies could be relaxed one year earlier,[10]

9. The estimate was obtained by estimating the direct first-round effect of lower oil prices on the CPI and WPI and then using a linearized reduced form of the wage and price equations of our econometric model to approximate subsequent effects.

10. The official discount rate was assumed to be reduced from 9.0 to 8.5 percent in April 1975 and then reduced further by 0.5 percent every three months through 1975.

and the rate of growth in real GNP could be improved as follows (in percent per fiscal year):

	1975	1976	1977
$10 price continued	2.1	4.2	7.9
Oil price falls to $6	3.4	4.7	8.7

The discussion of the past several pages may understate the problems facing Japan. A recent report of the Economic Planning Agency (EPA)[11] has concluded that the potential capacity growth of Japan's economy may well fall to less than 7 percent per year over the next five to ten years because of the inhibiting effects on growth-producing investment associated with higher-cost energy. In that case demand growth may have to be restrained for even longer periods than our econometric estimates imply.

Finally, while oil price increases are a significant cause of the current economic problems of Japan, those problems have other causes as well. The Japanese economy has been in the process of major economic and social transformation since 1968, a transformation characterized by shortages in labor supplies, slackening technological progress, and sharp differences in the population's views about the value of continued rapid growth. Quite independent of higher oil prices, therefore, Japan's economic growth is likely to be slower-paced and structurally different from that of the past two decades.

Appendix: Methodology and Assumptions

An input-output model was used to estimate the impact of a rise in the price of oil (from $3 to $9 a barrel, f.o.b. Persian Gulf) on the wholesale price of all commodities and on the prices of major commodity groups.[12] The calculations simulated the effect of a dollars-and-cents pass-through of higher oil costs but did not attempt to capture the secondary effects operating through induced changes in wage rates or in the prices of competitive fuels.

11. Economic Planning Agency, Government of Japan, "Possible Growth Path for 1975–1980" (in Japanese; mimeographed), December 1974.
12. The computational procedure used for this analysis is a somewhat modified version of the methods described by Tsunehiko Watanabe and Shuntaro Shishido in "Planning Applications of the Leontief Model in Japan," in A. P. Carter and Andrew Brody (eds.), *Input-Output Analysis* (Amsterdam: North-Holland Publishing Company, 1970), Vol. 2, pp. 9–23.

The basic price relationships can be expressed as follows:

(1) $P_{xi} = P_1 a_i + P_2 a_{2i} + \cdots + P_n a_{ni} + W_i L_i / X_i + S_i / X_i$

(2) $P_i = D_i P_{xi} + (1 - D_1) P_{qi}$

where

P_x = supply price of the ith sector
P_{xi} = price of domestic output in the ith sector
a_{ji} = input-output coefficient from the jth to the ith sector
P_{qi} = price of imported raw materials in the ith sector
D_i = weight of domestic output in relation to imported raw materials in the ith sector
L_i = labor input in the ith sector
W_i, S_i, X_i = wage rate, profits, and output, respectively, in the ith sector

A sixty-sector 1970 matrix was aggregated to a twenty-four-sector matrix, in groupings which emphasize the importance of various imported raw materials. Changes in wage rates, government controlled prices, profit rates, and prices of imported minerals, fuels, and all other imported commodities were considered as the exogenous variables. Values for these exogenous variables were assumed for each quarter of 1974 and annually for 1975 and 1976, on the basis of actual market experience, government changes in controlled prices, and expected future developments. Where it was relevant to do so, two sets of values were assigned to each exogenous variable, one assuming an oil price of $3 and the other a price of $9. Wage rates and profit rates, however, were not varied between the two cases.

The coefficients from the twenty-four-sector input-output model were then used to simulate two sets of supply price increases for each sector in 1974, 1975, and 1976. From the results of these simulations, Table 4-6 in the text was calculated, showing the *difference* between the two simulations—that is, the effect of varying only the price of oil from $3 to $9 a barrel, while maintaining the assumed values for nonoil exogenous variables. Table 4-7 shows the simulated increase in prices for each sector between 1972 and 1976 on the basis of the assumed changes in all the exogenous variables between 1972 and 1976, including a rise in the price of oil from its 1973 level to $9 a barrel.

CHAPTER FIVE

The Developing Countries

WOUTER TIMS

International Bank for Reconstruction and Development

AN ASSESSMENT of the consequences of the higher price of oil on the
developing countries and of the scope for possible policy responses cannot be made independently from a similar assessment for the other regions
of the world. The opportunities of the developing countries to take offsetting action are limited by their position in the world economy. Their
exports depend on markets in industrial countries where their products
compete with those from the temperate zones, and they cannot exert significant influence on the prices of their imports. Capital inflows finance a
substantial part of their imports and investments, and they have limited
access to international capital markets. The direct impact of the higher
oil price in raising import costs, therefore, represents only part of the
consequences: recession in the industrial countries reduces their exports
and foreign-exchange earnings and hence their capacity to import and
to support domestic investment, if there are no compensating additional
inflows of capital. Even when more foreign capital can be secured, the
cost of servicing it is an additional burden on the balance of payments and
on the resources that can be mobilized for investment in the future.

The short-run impact of higher oil prices on the developing countries
thus differs markedly from the consequences facing industrial countries.
Although aggregate demand is affected, as it is in the industrial countries,
the adverse consequences for the developing countries derive principally
from the direct and indirect impact of higher oil prices on their already
strained balance of payments. In short, their chronic difficulty in mobiliz-
ing scarce foreign-exchange resources to finance imports for development

169

is aggravated directly by the higher cost of the oil that they import and indirectly by the effect of higher oil prices in reducing economic growth and causing pressure on capital markets in industrial countries, which in turn weakens demand for the exports of developing countries and reduces their prospects for borrowing abroad on appropriate terms.

In order to present an analysis of the impact of higher oil prices on the developing countries with the least degree of entanglement, a distinction is made between the impact in the short run (1974–75) and in the longer run through 1980. The analysis also distinguishes, at least for the short run, between the direct impact of higher oil prices and the indirect effects through their trade and payment relations with the rest of the world, particularly with the countries belonging to the Organisation for Economic Co-operation and Development (OECD). Also, developing countries cannot be fruitfully analyzed as a monolithic group, since their structures, development levels, and international economic relations exhibit considerable diversity. The discussion therefore distinguishes separate groups of developing countries differing widely in the extent to which current economic events influence their prospects for development.

The cold numbers and analysis presented below mean that large numbers of people in the developing world face increasing misery and threats of tragedy. Even though these are not sketched out in any detail, it should be realized that they form the sorrowful leading theme of the analysis. The additional burdens inflicted upon the countries which are least capable of adjustment to the new economic realities reach far beyond the economic and financial consequences. Despite their inability to command adequate resources to meet their new problems, however, tragedy for these countries can be avoided. Their needs are not large in relation to the resources at the disposal of the richer nations and can therefore be met if there is sufficient will among the richer nations to assist the poorer ones. Given the prospects for many of the people in the developing world, the formulation of programs to stave off the negative impact of current developments should be undertaken soon if the increasingly common situation of assistance that is too little and too late is to be avoided.

Basic Assumptions

There are a number of variables of crucial importance to the developing countries that they cannot influence themselves to any significant degree. Notable among them is the future real growth of output and incomes in the industrial countries that are their main trading partners and are also the main source of their long-term capital inflows. Another is the rate of inflation in the world economy, since it affects the relative prices of the commodities produced in the developing countries for international trade. The third external factor—at present the most important—is the price of oil being set by the major oil exporters; it is not only the price level of oil, but also uncertainty about its future course, which constitutes a major problem for the oil-importing developing countries, as their capacity to import is critical to their investment plans. Finally, there is the need to make assumptions about the future availability of capital. This involves judgments regarding creditworthiness of developing countries and their access to international capital markets, as well as the estimation of future flows of concessional resources from both traditional and new donor sources. The extent to which concessional assistance may erode as a consequence of continued international inflation constitutes an element of additional uncertainty for those countries which depend heavily on such capital flows.

Growth in the OECD Countries

There do not exist any coherent projections of possible growth in the industrial countries except the ones assembled some years ago by the OECD Secretariat, which by now are no longer reliable. Neither the exceptionally high rates of growth in 1972 and 1973 nor the recession of 1974 and the probable stagnation in 1975 could be foreseen at the time of those forecasts, and the effect of the higher oil price on the growth of real income needs now to be taken into account as well.

The basic assumption made in those projections is that economic growth in the OECD countries for the full decade of the seventies will average 4.7 percent a year, which is a rate equal to that experienced in the preceding decade. Since the average through 1973 exceeded that rate by a substantial margin, a lower rate of growth between 1973 and 1980 is implied; most of that, however, would be centered in 1974 and 1975, which means that the second half of the current decade would again witness growth at approximately the same rates as the long-term trend. This projection is not without its controversial elements, for its basic implication is that the industrial countries are capable of adjusting to the higher

Table 5-1. *Past and Projected Growth of GNP in OECD Countries*
Annual average in percent

Country or region	1973	1974	1975	1975–80
North America	6.0	−1.7	−4.0	6.3
Western Europe	5.3	2.3	1.3	5.3
Japan, Oceania	9.5	−1.6	1.2	7.5
Total OECD	6.3	−0.1	−1.1	6.1

Sources: 1973–74 data and 1975 estimates from OECD; 1976–80 figures are IBRD staff projections.

price of oil and to its consequences in terms of financial flows and balance of payments financing with no more than a temporary loss of growth momentum (see Table 5-1).

The expectation of growth at long-term trend rates assumes that countries in the OECD group will adopt policies to maintain the rate of investment, consumer demand, and utilization of production capacity and employment, even if doing so entails high rates of inflation. It assumes further that these countries will continue to abstain from policies aimed at reducing their own balance-of-payments problems at the expense of other oil-importing countries. As time passes, however, it will become more difficult to distinguish oil-related deficits from the general balance-of-payments position.

Inflation

It is difficult to see how present rates of inflation can be reduced to the point of being no higher than those experienced during the sixties without either a significant reduction of growth and employment or major new policies in the principal industrial countries for controlling the processes which now determine nominal factor incomes. Neither alternative seems a likely possibility, nor can either be considered politically feasible. The average increase of GNP deflators in the OECD countries during 1973 was 7.5 percent, measured in national currencies (more than 13.5 percent when measured in U.S. dollars), and the increase for 1974 and 1975 will probably turn out to have been about 10 to 12 percent a year. A number of factors, including higher rates of unemployment, no further real (or perhaps even nominal) increases in the price of oil, and declining prices of other primary commodities point toward a reduction in rates of inflation in the course of 1975 and thereafter. A gradual decline between 1975 and 1980 is projected, with inflation still persisting at approximately 7 percent a year by the end of the decade.[1]

The Price of Oil

At present a long-term projection of the price of oil is difficult because of the unilateral character of the price-setting process during 1973 and 1974; in 1975 and beyond, market forces may assert themselves again, becoming stronger as time goes on, but this in itself adds to the uncertainty over future prices. The demand for imported oil, which in the long run will affect the price of oil increasingly, is a residual demand twice over: first, oil from all sources—not just from the members of OPEC—has become more expensive and may therefore cease to represent so large a share of the total consumption of energy as it does now, and second, a strong effort to increase oil production within the OECD countries and elsewhere will cut into the demand for oil produced by the members of OPEC. Small differences in estimates of the total demand for energy and the supplies of energy other than from OPEC sources can therefore lead to large differences in the estimated demand for OPEC oil. This uncertainty is compounded by confusion about the long-term supply policies of OPEC and of its member countries individually. In the short run, barring a major further downturn of economic activity in the OECD countries, the divergence between supply and demand for OPEC oil is not likely to get out of hand; there will thus continue to be some room for OPEC to set prices unilaterally, but OPEC's market power may decrease in the course of time.

The average f.o.b. price of Saudi Arabian light crude oil during 1974 was $9.78 a barrel; this implies an average f.o.b. price for all OPEC crudes of around $10.25 a barrel. This average includes an estimate of $10.46 a barrel for all OPEC crudes in the fourth quarter of 1974, and it is likely that OPEC will seek to maintain this price at least through most of 1975.

The depressed economic conditions in the OECD countries during 1974–75 and the short-run impact of higher oil prices reduced the consumption of oil and led to a reduction in the volume of OECD oil imports of about 5–6 percent from those of 1973. Imports for consumption in 1975 may fall below 1974 levels somewhat as real growth rates in the

1. Specific assumptions of the rate of inflation are 12 percent in 1975, 9.8 percent in 1976, 8.5 percent in 1977, 8.0 percent in 1978, 7.5 percent in 1979, and 7.0 percent in 1980.

Table 5-2. *Alternative Assumptions concerning Prices of Saudi Arabian Light Crude Oil, 1973–80*

Dollars per barrel, f.o.b. Persian Gulf

Alternative	1973	1974	1975	1977	1980
High					
In 1974 dollars	3.08	9.78	9.40	9.40	9.40
In current dollars	2.71	9.78	10.46	11.25	13.90
Low					
In 1974 dollars	3.08	9.78	9.40	8.75	7.00
In current dollars	2.71	9.78	10.46	10.46	10.46

import-dependent industrial countries fall further; the demand for imported oil for the building up of stocks, moreover, will probably decline. However, there continue to be constraints on the rapidity with which domestic production of primary sources of energy can increase. The members of OPEC are in a position to increase supplies of oil much faster than they have been doing, but they have restrained their output in the course of 1974 and early 1975 in order to maintain a reasonable balance with apparent demand and thus maintain prices. In future years a slow growth in the demand for OPEC oil may require the members of OPEC to allow more and more of their existing or potential production capacity to remain idle. If real prices should be maintained, the demand for their oil probably would grow scarcely at all from 1975 onward, and after 1980 an actual decline might set in as other sources of energy become available. The capacity of OPEC to maintain control over prices thus depends critically on the capacity of its members for joint control of their oil exports.

Alternative price projections are introduced here to reflect the uncertainties of the future oil market (see Table 5-2). The first assumes that the *real* price of oil will decline moderately in 1975 and then stabilize through 1980 at a price of $9.40 a barrel f.o.b. Persian Gulf for Saudi Arabian light measured in 1974 dollars. This is equivalent to a price of $13.90 a barrel in current 1980 dollars for that year. The lower alternative is based on the same assumption for 1975, but from then on it is assumed that the *nominal* price will be maintained, which means that the real price will erode through inflation in subsequent years; by 1980 the real price, measured in 1974 dollars, will have been reduced to about $7.00 a barrel for Saudi Arabian light f.o.b. Persian Gulf.

Capital Flows

Prospects for the nonoil-exporting developing countries depend significantly on their ability to attract capital flows of the level and terms appropriate to their potential for development and their capacity to service debt. In the analysis that follows, capital inflows for each country have been estimated on the basis of its access to capital markets; in addition, capital inflows for countries heavily dependent on concessional assistance are analyzed on the basis of alternative assumptions concerning the possible levels of contributions to these flows from the traditional and new donor countries.

The Short-Term Impact (1974 and 1975)

The increased oil price raises import costs to all developing countries that depend on imported oil to supply part of their energy. The magnitude of this impact depends on the level of development of each country, the degree to which it relies on imported oil for meeting its energy requirements, and the degree to which it participates in international trade.

The impact therefore varies particularly when the developing countries are grouped by income levels and by export structure. Countries exporting oil obviously constitute a distinct group of developing countries; similarly, those exporters of other minerals and metals that depend heavily on exports based on natural resources form another distinct group. The remaining developing countries are grouped into three categories, with per capita income boundaries at $200 and $375. The main characteristics of these country groups are shown in Table 5-3 below.

With respect to both the amounts of energy required and the kinds used, there are significant differences among these groups of countries. Countries at the lower end of the scale usually have small modern sectors and are lacking in infrastructure; these characteristics tend to be reflected in low levels of energy consumption per capita and by a composition of use dominated by traditional forms of energy and only limited amounts of oil. Energy consumption appears to rise over the scale of development levels at a rate faster than that of economic growth, and oil consumption, at least when the price of oil was comparatively low, rose faster than total energy consumption. The much higher levels of oil consumption at more

Table 5-3. *Principal Economic Characteristics of Three Groups of Developing Countries,*[a] *1968–72*

Group of countries	1972 population as percentage of total	Average annual percentage growth of GDP, 1968–72	Exports as percentage of GDP in 1972	Average annual growth of exports,[b] 1968–72, in percentage	Outstanding public debt per capita, 31 December 1972, in dollars
Oil-exporting countries	15.8	9.6	26.9	5.0	62
Other minerals exporters	3.7	4.8	21.9	4.6	119
Other developing countries according to per capita income:					
Above $375	21.2	7.1	11.1	7.5	94
$200–$375	9.3	6.6	19.9	10.2	95
Below $200	50.0	3.5	6.7	1.8	25
Total or average	100.0	6.2	14.2	6.4	55

Source: IBRD staff studies.

a. Data by group of countries in this and subsequent tables are based on a sample of forty-seven countries; the sample countries in each group, taken together, represent 90 percent or more of the entire group when it is measured according to population, GNP, capital inflows, and indebtedness. The number of sample countries in each group is given in parentheses.

Oil-exporting countries (7): Iran, Iraq, Venezuela, Nigeria, Algeria, Ecuador, Indonesia.
Other minerals exporters (7): Chile, Bolivia, Jamaica, Liberia, Morocco, Zaire, Zambia.
Countries with per capita income above $375 (12): Greece, Tunisia, Yugoslavia, Argentina, Brazil, Colombia, Dominican Republic, Guatemala, Mexico, Peru, Uruguay, Malaysia.
Countries with per capita income between $200 and $375 (11): Egypt, Syria, Turkey, Korea, Philippines, Thailand, Cameroon, Ghana, Ivory Coast, Senegal, Sierra Leone.
Countries with per capita income below $200 (10): Ethiopia, Kenya, Mali, Sudan, Tanzania, Uganda, Bangladesh, India, Pakistan, Sri Lanka.

b. In constant 1967–69 prices.

advanced levels of development constitute in most cases a claim on foreign-exchange resources, since most of the oil used in these countries must be imported. This does not necessarily imply, however, that oil imports constitute a heavier burden on developing countries at the upper end of the income scale, since participation in international trade also increases with higher income levels. The lower-income countries may in fact be affected more seriously by burdens on their foreign-exchange resources, since their exports, and hence their capacity to finance imports, are small in relation to their national product. Significant statistics related to energy consumption by these three groups of countries are given in Table 5-4.

Oil-importing developing countries imported an estimated 192 million tons in 1973, or 13 percent of world oil trade. In 1973 prices the cost of

Table 5-4. *Energy Characteristics of Three Groups of Developing Countries, 1972*

Group of countries	Energy consumption per capita (kilograms of coal equivalent)	Imported oil as percentage of energy consumption	Imported oil as percentage of total imports (goods plus non-factor services)
Oil-exporting countries	567	. . .	. . .
Other minerals-producing countries	602	42	4
Other developing countries according to per capita income:			
Above $375	1,021	38	5
$200–$375	402	45	5
Below $200	149	27	8
Average for all countries	410	31	4
Average excluding oil-exporting countries	390	37	6

Source: U.N. Energy Statistics, Series J; IBRD Economic and Social Data Bank.

these imports was about $4.2 billion; in 1974 prices, however, the cost of the same volume was about $15.2 billion, or an increase of $11 billion in one year. Imports of goods in 1973 by these developing countries that import oil are estimated at about $82 billion; thus the higher price of oil alone increased the cost of imports in 1974 by about 13 percent over those of 1973.

Prices of other primary commodities are projected by the International Bank for Reconstruction and Development (IBRD) on the basis of prospects for demand and supply and sensitivity to international inflation. These projections, summarized in Table 5-5 below, indicate that the high prices of 1974 are likely to have constituted a peak, to be followed by a decline in relation to international purchasing power in 1975. In the long run, prices are seen to return to the same level as in the late sixties in terms of purchasing power.

To assess the full effect of price changes on the trade position of the developing countries, the additional cost of oil imports should be combined with estimates of the costs of increases in the prices of their other imports; then the higher prices realized by these countries from their exports must be set off against the total effect of import price increases. The net results for each group of countries are quite different, as shown in Table 5-6.

Table 5-5. *Projected Prices of Primary Commodities in International Trade, 1972–80*

1967–69 = 100; deflated by the export price of manufactured goods

Commodity group	1972	1973	1974	1975	1980
Major primary commodities[a]	95	118	235	227	220
Excluding petroleum	85	106	122	106	94
Agricultural products	90	110	127	111	94
Food	91	106	136	122	95
Nonfood	85	122	107	84	93
Metals and minerals	74	92	106	89	91
International price index[b]	127	150	182	202	294

Source: IBRD staff projections.
a. Thirty-five commodities are included in this group.
b. Weighted average export price of manufactured goods from major industrial countries; used as deflator for current-dollar prices underlying the indexes for primary products shown above.

The additional cost of oil for developing countries is larger than the total cost of the food grains that they import. Rising prices of food grains did, however, have their effect earlier: in 1973 the costs of imported food grains had risen some $4.5 billion above the level of the preceding year, largely as a consequence of price increases. The fact that the prices of food grains stayed at these high levels in 1974—in fact, they rose somewhat further on an average—implies that the higher oil price came on top of the already heavy financial outlays for needed food grains.

On the basis of the price estimates and projections in Table 5-5, one can infer for each group of countries the costs and gains in 1974 and 1975 by expressing the export and import volumes of 1973 in prices of 1974

Table 5-6. *Price Changes in International Trade of Developing Countries, 1973–75*

1967–69 = 100

Group of countries	Export prices			Import prices			Terms of trade		
	1973	1974	1975	1973	1974	1975	1973	1974	1975
Oil-exporting countries	203	557	600	152	178	195	134	313	308
Other minerals exporters	140	179	167	147	199	213	95	90	79
Other developing countries according to per capita income:									
Above $375	146	195	194	140	201	215	104	97	90
$200–$375	143	188	190	142	201	214	101	94	88
Below $200	141	174	180	144	207	216	99	84	83
Total, excluding oil-exporting countries	145	190	191	142	202	215	102	94	89

Source: IBRD staff projections.

Table 5-7. *Gains and Losses from Terms of Trade, 1973–75*
Billions of dollars

Group of countries or commodities	1973		1973 volumes at 1974 prices		1973 volumes at 1975 prices	
	Exports	Imports	Exports	Imports	Exports	Imports
By country group						
Oil exporters	41.5	21.9	118.8	25.7	126.5	28.0
Other minerals exporters	8.3	5.2	10.6	7.0	9.9	7.5
Other developing countries						
Above $375	39.6	46.3	52.9	66.5	52.6	71.1
$200–$375	13.6	16.4	17.8	23.2	18.1	24.7
Below $200	9.2	9.9	11.4	14.2	11.7	14.9
By commodity group						
Oil and products[a]	42.7	7.1	122.0	24.0	129.5	25.2
Food grains	2.6	8.3	3.7	11.0	3.3	10.5
Other foods, raw materials	43.1	24.3	56.9	25.8	54.3	27.0
Manufactures	23.8	60.0	28.9	75.8	31.7	83.5
Total	112.2	99.7	211.5	136.6	218.8	146.2
Total excluding oil exporters	70.7	77.8	92.7	110.9	92.3	118.2

Source: IBRD staff projections.
a. Import figures are significantly higher than those quoted earlier for imports of net importers alone. This is because of the inclusion of countries which are substantial importers of crude oil and exporters of refined products; the numbers therefore are offset by a higher export figure.

and 1975. Specific elements such as oil and food grains are shown separately, in order to give an impression of the relative importance of price changes for those commodities in comparison to the overall effect of prices on the values of imports and exports of each group of countries (Table 5-7).

Earlier it was shown that in 1974 the oil-importing developing countries had to spend $11 billion more than they did in 1973 for the same quantity of oil. Table 5-7, which includes also those developing countries that are minor exporters of oil and those that both import and export oil and petroleum products, shows a cost increase that is considerably larger because of this coverage. As a result of actual and projected price increases, the cost of other imports increased $20 billion in 1974 and $28.4 billion in 1975 over the 1973 level. On the other hand, earnings from exports other than oil in both 1974 and 1975 are about $20 billion higher than they were in 1973. Thus, the picture can be summarized as follows (in billions of dollars):

	1974 over 1973	1975 over 1973
Increase in the net cost of oil	11.0	11.0
Increase in the cost of other primary products	4.2	4.9
Increase in the cost of manufactures	15.8	23.5
Increase in export earnings	20.0	19.8
Net change in trade position	−11.0	−19.6

In overall terms, the developing countries appear to gain substantially from the price developments of 1974, and they will continue to realize large gains in 1975: their trade surplus of $12.5 billion in 1973 prices rises, for the same volumes of transactions, to $75 billion in 1974, and it is expected to be some $73 billion in 1975. These figures, however, present a misleading picture. The net gain is not just located within one group of countries, but in fact consists of very large gains accruing entirely to the oil-exporting countries and rising balance-of-payments deficits on the part of all other developing countries except the exporters of other minerals, which appear to be just holding their own.

As is shown by way of summary in Table 5-8, the oil-producing countries gain between $70 billion and $80 billion a year, whereas the other developing countries lose an average of about $13 billion a year. This amount is larger than the additional cost of oil imports to them; in other words, other price changes in both imports and exports also contributed to a decline in their position, at least during these two years.

The deterioration in the position of the oil-importing countries, as calculated from the 1973 volume of trade, does not take into account the impact of changes in the volume of trade in 1974 and 1975 or of changes in factor and nonfactor services. On both counts a decline in the current

Table 5-8. *1973 Balance of Trade Expressed in 1974 and 1975 Prices*
Billions of U.S. dollars

Year	Oil exporters	Other minerals exporters	Other developing countries by income group		
			Above $375	$200–$375	Below $200
1973	+19.6	+3.1	−6.7	−2.8	−0.7
1974	+93.1	+3.6	−13.6	−5.4	−2.8
1975	+98.5	+2.4	−18.5	−6.6	−3.2

Source: Derived from Table 5-7.

Table 5-9. *Balance of Payments of Oil-Importing Developing Countries, 1973–75*

Item	1973	1974	1975
Exports of goods, f.o.b.	70.7	101.7	92.7
Imports of goods, c.i.f.	77.8	128.5	128.8
Trade balance	−7.1	−26.8	−26.1
Nonfactor-service receipts (net)	−2.6	−3.1	−3.4
Resource balance	−9.7	−29.9	−29.5
Factor-service receipts (net)	−0.7	−1.0	−2.3
Current-account balance	−10.4	−30.9	−31.8
Net medium and long-term capital	22.4	30.7	33.4
Changes in reserves and short-term capital movements (increases [−], decreases [+])	−12.0	0.2	−1.6

Source: IBRD staff projections.

balance emerges. A detailed country-by-country projection of balances of payments for 1974 and 1975 suggests that the increase of the current-account deficit over that of 1973 amounts to some $20 billion in each of these two years (see Table 5-9). Somewhat more than half this amount can be attributed to the increase in the cost of oil imports resulting from higher oil prices.

Since the current-account deficit was unusually small in 1973, net capital inflows proved to be large enough both to finance the gap and to permit an increase of about $10 billion in foreign-exchange reserves. The tripling of the current-account deficit in 1974 would have undone this reserve gain entirely, had the developing countries not been able to attract and absorb an additional $8.3 billion net in medium- and long-term capital. Although some countries lost reserves, these countries as a group were able to maintain their reserve levels; still, inflation as reflected in international trade prices reduced reserves as a percentage of imports significantly.

A substantial part of additional capital inflows in 1974 and 1975 originate in the oil-exporting countries. Bilateral contributions and flows through various multilateral agencies in 1974 and 1975 may amount to $7 billion to $8 billion on an average, compared to $1.2 billion in 1973.

In the short run the developing countries are thus faced by a major deterioration in their terms of trade, in part a direct result of the higher price of oil, in part owing to the effects of the recession in the industrial countries.

The middle- and higher-income countries that have diversified exports are vulnerable to fluctuations of growth in the industrial countries. Prices of their exports may not be affected so strongly as are those of the lower-income countries, but fluctuations in the volume of demand for their exports are quite pronounced. This suggests also that for the middle- and higher-income countries the restoration of normal growth in the OECD countries is of prime importance as the factor that can quickly restore the level and growth of their export earnings. For the lower-income countries, however, the principal effect of higher oil prices and the OECD recession is a major deterioration of their terms of trade, which are not likely to recover at any time in the next five years.

The capacity of the low-income countries to adjust in the short run is strictly limited. Energy uses cannot easily be reduced without affecting output. Exports, consisting mainly of primary commodities, cannot be increased rapidly, particularly when international demand is weak. Unless adequate net inflows of capital (including the possibility of reduced debt-service payments) can be generated, imports must be held down, thus in turn affecting the growth of output and incomes. Additional capital flows can take a variety of forms, but they must in any event be geared to the immediate import-financing needs of the developing countries. Program aid is therefore preferable to project financing, and rescheduling of debt-service obligations is, for debtor countries, the most attractive alternative. Within capital flows of the program type, those associated with food and with oil need to be mentioned specifically, for the additional capital flows in these forms can be tied to the commodities that account in large measure for the balance-of-payments problems of the developing countries.

In summary, the extent to which developing countries can meet their short-term problems through a reallocation of resources is limited. This is particularly the case in countries at the lower end of the income scale, the economic structures of which are usually characterized by a lack of diversification. The difficulties of adding to net capital inflows of all developing countries—but especially of the lower-income group—tend to make plain the benefits to these countries of accelerated growth in the industrial countries, which constitute not only the main source of development capital, but also the main outlet for the exports of the developing countries. Conditions should improve in 1976 and 1977, if the recovery of the industrial countries is substantial and international trade resumes its growth. Many of the developing countries will, however, not return to their original growth path, for they are living with reduced levels of invest-ment in 1974 and 1975, and their production capacity will thus grow less in subsequent years. Those countries that borrow heavily to maintain growth in the short run take the risk of reducing their ability to borrow later and thus of becoming more vulnerable to external economic shocks in the future.

The preceding analysis of balance-of-payments prospects for the de-veloping countries is based on an assumed maintenance of the oil price at $9.40 a barrel for the "marker" crude, Saudi Arabian light, in 1974 dol-lars. If the alternative price projection were to be realized (see Table 5-2) and the oil price were to decline gradually after 1975 to $7.00 a barrel in 1980, the outlook for the developing countries, particularly in the next two years, would not change markedly. In 1977 the price difference (in current dollars) would amount to just about $2.00 a barrel, or a reduction of import costs by some $4 billion in that year. This would reduce net capital requirements of the oil-importing countries by some 10–15 per-cent and would be particularly significant for the lower-income countries.

Long-Term Prospects

In the long run, consideration must be given to the various possibilities for energy substitution, both by switching from imported oil to other pri-mary sources of energy for existing facilities and by investment in new facilities that generate or use alternative forms of energy. In the short and medium term—that is, through the remainder of the present decade—substitution possibilities in the developing countries are limited. In the power sector, where internal adjustments are the most significant, long lead times for new investments will reduce the scope for substitution over this period; it is only in countries with large power systems that substitu-tion of other primary sources of energy (notably coal) for oil can reduce the share of oil in energy consumption.

It is not likely, therefore, that the higher cost of imported oil will be offset to any great extent by either conservation or substitution. Invest-ment programs in the developing countries will need to be adjusted so that they can accommodate substantial new programs and projects for the future production of domestic energy. Such investments, besides entailing long lead times, also tend to be highly capital-intensive and to require sub-

stantial imports of machinery and equipment. Before any reduction in
foreign expenditures for oil imports can be realized, these countries must
increase their foreign expenditures further in relation to their adjusted investment needs; alternatively, they must reduce their imports of goods and
services other than oil, whereas most of such imports are directly associated with the development programs of these countries and with the maintenance of stable prices on major items of domestic consumption.

It is unlikely that the adjustments in the investment programs of the
developing countries can be made by adding the new investment needs to
existing development programs. In fact, if capital flows do not increase
sufficiently to maintain the volume of imports in relation to a country's
growth objectives, development will slow down, and generation of domestic resources for investment will inevitably be reduced; even maintaining
planned levels of investment will therefore be difficult, and the accommodation of capital-intensive energy-related projects will further reduce
the resources available for the development of other sectors of the economy. Countries are thus faced with painful choices regarding allocation of
the cuts in investments outside the energy sector. One can speculate that
the sectors with the least institutional strength in defending their shares of
development resources will suffer most, and one may thus expect that the
social sectors—education, health, nutrition—and new activities, such as
integrated rural development, will be hurt more than infrastructure investment, such as modern transport and industry. It is too early to cite any
significant examples; it must be borne in mind, moreover, that adjustments
in investment programs are not immediately visible. Ongoing projects
absorb by far the major part of annual development programs of most
countries, and substitution between projects and sectors can be done only
at the margin of the program; a shortfall of development finance resulting
from reduced growth and higher project costs can in itself lead to further
delays, because ongoing projects then take longer to complete, thereby
postponing the time at which energy-substitution projects can be undertaken.

The oil price increase has caused a number of previously unattractive
energy-substitution projects—hydroelectric schemes, coal developments,
geothermal energy, and generation of nuclear power—to become economically feasible. In the field of nuclear energy, recent developments
make it possible to build smaller plants which can still compete with oil-

fired generating plants; a number of countries whose medium-term energy demand prospects were in the past considered too small for nuclear power now find that this type of energy is an economic possibility.

Most equipment and machinery required for energy substitution needs to be procured in the industrial countries, which are embarking on similar programs themselves. Developing countries may therefore experience difficulties in purchasing and taking delivery of such equipment, for they face competition from virtually all other countries. Industrial countries will tend to place large orders, generally consisting of large units, for series of equipment deliveries; the orders of the developing countries are usually for single items and smaller units, preferably financed by the supplier by means of concessional credits. This situation, added to the problems of financing energy-substitution projects, may increase the time required for significant progress to be made in reducing dependence on imported oil.

In view of all these factors, the developing countries are likely to reduce their dependence on imported oil slowly at best, lagging significantly behind the industrial countries in this respect. Of course, some countries that are in a position to develop domestic sources of energy, such as coal, oil, and gas, will do so; thus, not all countries will remain as dependent on imports as at present. Therefore, oil imports which in the sixties increased by 9–10 percent a year on an average are expected to grow at a reduced rate in the years to come. Still, with the projected rate of growth of income in the developing countries, it appears unlikely that their oil imports could grow in volume by less than 5–6 percent a year between 1973 and 1985. All this must be taken into account in the assessment of the long-term outlook for the developing countries; reasonable rates of growth in line with actual performance through 1973 can only be achieved with the assistance of large additional foreign exchange resources, both to meet the cost of imported oil and to finance the adjustment investments intended to reduce dependence on imported oil over the long term.

Additional foreign-exchange resources can be provided in a variety of ways: increased earnings through producers' arrangements that prove to be successful in raising prices of exports of raw materials; improved access to overseas markets, particularly for exports of manufactured goods; additional flows of external capital in the form of grants or loans; or reductions of the cost of debt-servicing through renegotiation of such liabilities. In the short run the use of reserves and short-term borrowing may to some extent bridge the gap for countries to which such means are available; in the long run these cannot be relied upon to any significant degree as a way of meeting the additional foreign-exchange costs.

The developing countries earn more than 80 percent of their foreign exchange (excluding net capital flows) from exports to the industrial countries. The long-term prospects for these earnings are therefore almost entirely a function of demand and price developments in the industrial countries. Supply prospects for primary commodities, which constitute a major part of the exports of the developing countries, can play an important role, particularly when the products concerned are those that developing countries supply in competition with natural or synthetic products from the industrial countries. The trade policies adopted by the producers or the consumers of these export goods add another element of uncertainty to those associated with demand, supply, and the effects of international inflation.

The volume of exports of the developing countries (including oil) increased at a rate of about 6.5 percent a year during the sixties; this rate accelerated to about 7.5 percent in the early seventies and is estimated to have been 8 percent in 1973. The most rapid growth occurred in the export of manufactured goods, followed by nonfactor services and petroleum and products.

If the OECD countries should maintain an average growth rate of 4.1 percent a year between 1973 and 1980 and the oil price should be maintained in real terms at the level projected for 1975 of $9.40 a barrel for marker crude, export volumes of the developing countries as a group could be maintained on a growth path of about 7.5 percent a year, as they have during the last five years. This figure includes exports of oil, which are projected to grow by only 1.2 percent a year; other exports of developing countries are projected to grow more rapidly than in the past, reflecting in part the increasing importance of services and manufactures in their exports, but also reflecting higher projected rates of growth for agricultural products.

The export growth rates projected in Table 5-10 are predicated upon the assumed rapid recovery of the OECD countries from the low level of economic activity in 1975. If the period of recovery is longer, and full employment is not reached until after 1980, average growth rates for OECD countries could be reduced by a full percentage point a year, averaging some 5 percent a year between 1975 and 1980. Exports of the

Table 5-10. *Actual and Projected Growth of Exports of Developing Countries, 1967/69–1980*

Average annual percentage rates

Commodity group	1967/69–1972/74	1973–1980
Food, food products	1.6	3.2
Agricultural products other than food	2.6	3.6
Metals, minerals, and ores	5.4	5.5
Petroleum and products	9.5	1.2
Manufactured goods	n.a.	13.2
Services	n.a.	10.4

Source: IBRD staff estimates and projections.
n.a. Not available.

developing countries are quite sensitive to the growth performance of OECD countries, showing a reduced growth rate of their export volume of about 1.5 percent for each percentage point reduction of GNP growth in the OECD countries. Higher-income countries with relatively larger exports of income-elastic manufactured products appear to be somewhat more exposed to such influence than the lower-income countries. In addition to the volume effect of lower OECD growth, there is also a significant terms-of-trade effect amounting to some 0.2 percent deterioration per 1.0 percent reduction of OECD growth. A reduction of the oil price, as suggested in the low alternative of Table 5-2, amounting to a reduction of 25 percent in 1974 dollars between 1975 and 1980, would not have any noticeable effect on export volumes of the oil-importing developing countries but would improve their terms of trade by about 0.6 percent a year.

In this assessment, the good prospects for fats and oils, food grains, timber products, and cotton stand out over the less favorable outlook for sugar,[2] fruits and vegetables, and beef. The composition of exports in each group of countries is different, and the anticipated volume of trade by group of countries is therefore affected by the differences in the prospects for the various commodities. It is possible that growth in the volume of exports for the higher- and middle-income countries will continue through 1980 at rates comparable to those of the past five years; oil-exporting countries will experience a considerable slowdown of growth in the volume of exports, while exporters of other minerals are expected to achieve some small gains in volume. The lower-income countries, however, will experience a more favorable growth in export volumes. The

2. Prices are expected to decline from present high levels to a level more nearly comparable to those prevailing in the early seventies.

Table 5-11. *Export Volume and Price Prospects, 1968–80*

Average annual percentage rate of change

Group of countries	Export volume		Export price		Export value	
	1968–73	1973–80	1968–73	1973–80	1968–73	1973–80
Oil-exporting countries	9.0	4.0	15.8	25.8	26.2	30.8
Other minerals exporters	5.4	6.8	9.2	10.1	15.1	17.6
Other developing countries according to per capita income:						
Above $375	7.3	7.4	9.7	9.0	17.7	17.1
$200–$375	10.3	9.2	7.9	8.7	19.0	18.7
Below $200	1.9	5.4	7.1	7.6	9.1	13.4
Total, all developing countries	7.4	6.6	11.2	14.9	19.4	22.5
Total excluding oil-exporting countries	6.9	7.5	8.7	9.0	16.2	17.2

Source: IBRD staff estimates and projections.

projected rate of growth is more than 5 percent a year, compared with about 2 percent a year over the past five years. Exports of manufactured goods are a large part of the reason for this more optimistic outlook.

Price projections for exports of primary commodities are hazardous in a situation characterized by large price movements, uncertainties about markets and supplies, and unprecedented rates of international inflation. An analysis of the major primary commodity markets and their prospects leads to the conclusion that relative prices (that is, relative to general inflation) of almost all primary commodities except oil will return from their high levels of 1973 and 1974 to the average level of the sixties before the end of the current decade. The price forecasts translated into projected rates of growth of export prices by country group are presented in Table 5-11.

The projected export price increase for the oil-importing developing countries of 9 percent a year between 1973 and 1980 must be compared with an import price increase of slightly more than 10 percent a year. Thus, for the period as a whole, the terms of trade show a deterioration, substantial in 1974 and 1975, with a gradual recovery thereafter.

Changes in the trade regimes which at present govern the flows of agricultural commodities and the semimanufactured or manufactured products based on them could be of benefit to the developing countries that export these commodities. Complete liberalization of trade (abolishing duties and quantitative restrictions, including those embodied in the present agricultural policy of the EC) by the OECD countries for major primary commodities would increase the volume of exports and, in some instances, export prices as well. By 1980, earnings from the major primary commodities affected by import liberalization could be approximately 25 percent a year more than they would have been otherwise, and total export earnings of the oil-importing developing countries during the period 1973–80 might be increased by 1 percentage point a year over the rates shown in Table 5-11.[3] These estimates rest, however, on optimistic assumptions concerning the possibility of more rapid growth in export supplies from the developing countries; in the past, supply responses to improved market opportunities have not been impressive. Liberalization of trade restrictions on manufactured goods will be of increasing importance over the coming years and is of particular significance for a group of developing countries that have already made progress in promoting export sales of manufactured goods. The adjustments in the international division of labor associated with more liberal trade policies for manufactures are difficult to achieve; much will depend on the outcome of the current multilateral trade negotiations and on the willingness of governments to use import liberalization as a means of combatting domestic inflation.

Following the success of the action by OPEC to raise oil prices, developing countries are exploring the possibilities for similar action to restrict output and raise prices of other primary commodities. The effectiveness of such action will depend on the nature of the demand for a given commodity, the structure of its market, the feasibility of common action, and the scope for long-run substitution from other sources. If producer co-operation is limited to the developing countries, few commodities meet the requisite conditions for restricted output and consequent higher prices, and the increases in earnings to be achieved by these means will be relatively small, although important to the producing countries. Where large investments are needed to expand world production outside the group of co-operating countries, the resulting benefits to the members of the producer's cartel might be substantial for limited periods, particularly if simultaneous action were taken with respect to a number of commodities that are close or partial substitutes for each other.[4]

3. The basis for this estimate is "Possible Effects of Trade Liberalization for Primary Commodities," IBRD Working Paper No. 193 (January 1975).

4. A fuller discussion of these issues in the light of the most recent experience can

Capital Requirements

Net flows of long-term loans and grants from all sources to developing countries amounted to some $31 billion in 1973 (see Table 5-12).

Out of the total net flow of loan and grant capital of $31 billion in 1973, almost $3 billion was directed to the oil-exporting countries; the total also includes substantial amounts of capital representing the estimated value of technical assistance and some capital not identified by country of destination. The latter two account in 1973 for a registered amount of $5.6 billion. Identifiable flows of capital (net) that were made available to oil-importing developing countries for import financing thus amounted to $22.4 billion in 1973 and rose to an estimated average of $32 billion in 1974 and 1975 (see Table 5-9).

Table 5-12. *Net Capital Flows to Developing Countries in 1973*
Billions of U.S. dollars

Group of countries	Total	Net official capital	Net private loans	Direct private investment
Development Assistance Committee (DAC) countries	28.1	12.3	9.1[a]	6.7
Socialist countries	1.5	1.5	. . .	. . .
Oil-exporting countries	1.2	1.2	. . .	. . .
Total	30.8	15.0	9.1	6.7

Sources: Statistical Annex to the Report of the Chairman of the Development Assistance Committee of the OECD for 1974; capital flows recorded by recipient countries from sources other than DAC member countries; information from OECD, the UN Emergency Office and IBRD (partly collected from press reports) on capital flows originating in member countries of OPEC. All figures shown are net of (estimated) amortization payments and exclude contributions to multilateral agencies, but include net capital flows from those agencies to developing countries. Receipts by developing countries are therefore smaller in 1973 by about $70 million due to the net contributions of capital exporting countries to multilateral agencies.
a. Includes borrowing in international capital markets.

These amounts are in fact smaller than will be needed in the long run, for terms of trade will continue to be unfavorable after 1975, and foreign capital requirements will therefore increase further, from about $32 billion in 1974 and 1975 to between $70 and $75 billion by the end of the present decade, if the GNP growth objectives of the developing countries

be found in *Trade in Primary Commodities: Conflict or Cooperation?* A Tripartite Report by Fifteen Economists from Japan, the European Community, and North America (Brookings Institution, 1974). The limited scope for action by producers is also alluded to in a paper by Ernest Stern and Wouter Tims, "Changes in the Bargaining Strengths of the Developing Countries" (processed).

Table 5-13. *Net Capital Requirements for a Growth Rate of 6 Percent, 1975–80*

Billions of current U.S. dollars

Source	1973 actual	Average 1974/75 estimated	Average 1976–80 projected available	Average 1976–80 projected required
Net official capital	10.5	15.5	24.0	...
Net private capital	11.9	16.5	18.5	...
Total	22.4	32.0	42.5	70–75

Source: IBRD staff projections.

are to be maintained at an average of 6 percent. It should be noted, however, that all these figures are measured in current dollars and should be adjusted for a considerable element of inflation. In constant 1974 dollars, capital requirements by 1980 would be $50 billion to $53 billion. Against these requirements, present estimates suggest that only $33 billion (in 1974 dollars) will be available, or little more than is available at present. These estimates are based on known intentions of donors and estimated limits of creditworthiness for private capital. It is useful to present projections in current prices also, as in Table 5-13, because commitments, which precede net disbursements by several years, need to be increased substantially in the early years in order to achieve the high nominal disbursement levels that will be required toward the end of the decade. This is particularly true for the roughly 50 percent of these flows that originate in the public accounts of the creditor countries, as those commitments tend to lag behind inflation and are normally fixed in nominal terms for budget purposes. Therefore, instead of being geared to the high nominal disbursement needs in future years, these amounts are continuously eroded by current inflation.

The amount of capital required over and above the amount available is so large that serious analysis is hardly required to demonstrate the impossibility of attracting and absorbing such large inflows without impairing creditworthiness and creating excessive debt-service burdens. The conclusion inevitably follows that a growth rate of 6 percent between 1975 and 1980 for the oil-importing developing countries is not possible if it is to be achieved through capital flows alone. The rate of growth that is possible for these countries to achieve, given the amounts of capital available to them and the terms of its availability—and assuming no major changes in their own policies—is probably between 4 and 5 percent a year, depending on the speed of economic recovery in the OECD countries. And even under the most optimistic assumptions regarding growth in the industrial countries, it would still be a major achievement if the developing countries were to return gradually to a rate of growth again reaching 6 percent toward the end of the decade. Achievement of such an aim by 1980 may still require additional capital flows of around $8 billion a year (in 1974 dollars) between 1976 and 1980.

The alternative of increasing exports to permit a more rapid growth of the capacity to import is difficult to bring about not only because of its implications for savings and investment in the developing countries, but also because of the high rates of growth of exports already assumed.

The higher-income developing countries, among which are the heaviest borrowers, would need to expand their exports of manufactured goods to the industrial countries. Growth in the OECD countries is therefore of crucial significance for these countries. But even if they could export more than had been assumed, they would still need better access to international capital markets, where at present they must compete with many of the industrial countries also needing to finance large balance-of-payments deficits through international borrowing. This is the more important in view of the fact that a number of these higher-income developing countries still receive substantial amounts of concessional capital; pressure to switch such lending to countries which depend entirely on such funds will gradually increase, thereby increasing the amounts that the higher income countries need to raise by borrowing at market terms.

The higher- and middle-income countries are therefore strongly interested parties to the discussions of new mechanisms for the recycling of oil surpluses. Institutionalization of some of these flows—through the IMF oil facility, for example—affords them easier access to capital, since allocations will be made according to need, rather than on the basis of competitive strength or judgments of current and prospective creditworthiness. In this respect these developing countries have concerns in common with several of the industrial countries that are at present experiencing large balance-of-paments deficits for the financing of which they must have recourse to the international capital market. Also, many of these countries may have the basic resources for low-cost energy substitution, and they should therefore receive additional capital inflows for the development of these resources, both for domestic use and for export to other countries having oil deficits.

The poorest countries constitute the most vexing problem of all, since by far the largest part of their resources must be provided on the softest possible terms. Their net capital inflow in 1973 amounted to $4.4 billion; an addition of about 50 percent to that flow by 1976 would be required just to offset their terms-of-trade losses, of which more than half are derived from oil alone; it still would not allow these countries growth in GNP at any higher rate than some 3 percent a year, which barely exceeds the rate of their growth in population. In order to achieve a growth rate of 4.5 percent a year—still low in relation to needs—the net inflow would have to rise to about $13 billion by the end of the decade. Even so, if one considers the nominal GNP projected through 1980 for the members of the DAC in the aggregate, it becomes apparent that providing capital inflows of this volume would only require a maintenance of the present share of their total income represented by concessional aid. Doubts about the possibility of achieving even such a modest target stem from the fact that this share has been declining for some years already and from the expectation that it will be difficult to maintain even the real value of commitments in a period of inflation, given the way in which the governments of donor countries draw up and implement their budgets. Capital flows from OPEC brought about a significant increase in total flows to developing countries in 1974, but those are bound to taper off in later years.

In view of the situation described above, the need to reallocate the amounts of concessional assistance being made available at present becomes imperative, and the issues of access to international capital markets and the provision of capital on intermediate terms therefore gain further in importance as possible means of bringing about a significant shift of concessional capital to the poorest countries.

It is not necessary that all the additional concessional capital required come from countries that are now members of the DAC: several of the members of OPEC took major initiatives during 1974 to provide concessional resources to the poorest countries. The commitments that they made in 1974 are estimated to be $18 billion, of which about $6 billion was disbursed in 1974. This level of commitments is not likely to continue, but disbursements in the next few years will rise, then decline again toward the end of the decade. This will alleviate the capital requirements of the next few years, but in subsequent years it will necessitate substantial further increases in commitments from other sources—notably from the countries of the DAC—if the flow of real resources to developing countries is to grow at a positive rate.

Nothing has been said above about the domestic policies of the developing countries that might alleviate the foreign-exchange constraint on development; little is known at this time about the changes they themselves may induce to expand their export earnings, accelerate import substitution, and increase savings in order to maintain their investment levels. There can be no doubt that many of the developing countries, particularly in the higher-income group, do have some flexibility to adjust and to accelerate their growth in the next few years with lesser capital inflows than the "requirements" projected above. The degree of flexibility is related to the extent of diversification of their economies, the size of their own domestic markets, and the decisiveness of their governments in promoting the adjustment measures considered necessary, even if unpopular. That is a problem faced by all governments that need to meet the challenge of the balance-of-payments distortions that have followed the oil price increase.

Conclusions

Higher oil prices have had a substantial adverse effect on the economic prospects of the developing countries. They contributed heavily to a reduction in the terms of trade of these countries and to a long-lasting impairment of their capacity to import. These consequences have been aggravated recently by the effect of the recession in the industrial countries, which put downward pressure on both the volume and the prices of the exports of developing countries.

In the short run most developing countries were able to maintain the level of their imports and their growth during 1974; their exports were still growing, prices on the average were high, and their current-account position, which was fairly strong in 1973, could suffer some deterioration without constraining their economic progress. In 1975 the situation has become more difficult in a number of respects. Terms of trade continue to become less favorable, capital flows are being eroded by international inflation, and reserves in an increasing number of countries are becoming too small to be drawn down any further without the acceptance of serious risks. This situation prevails particularly with respect to countries that depend on borrowing in the international capital market and must therefore maintain adequate reserves in order to demonstrate their creditworthiness.

Economic prospects for the years beyond 1975 depend heavily on the rate of recovery in the OECD countries. Under favorable conditions, the higher-income countries should be able to restore their growth momentum, since their situation depends more on the growth of external markets than on their terms of trade. Further, they have significant flexibility to adjust by the use of domestic policies which promote both exports and import substitution and stimulate more efficient use of scarce resources. A judicious use of external borrowing, particularly in the early years, may be required to maintain growth rates before the effects of recovery in the OECD countries begin to be felt. Were that recovery to be slower, these countries could face serious creditworthiness problems associated with continued high levels of borrowing, an alternative to further reductions in growth of production and incomes.

The middle-income countries will have more difficulties maintaining their growth, since their borrowing capacity is more constricted and domestic adjustments are more difficult because of the relatively small size of the modern sectors in their economies. Although in principle facing the same challenges as the higher-income countries, their staying power is clearly less, and downward adjustments may therefore come earlier and be more severe.

For the poorest group of countries the most likely outlook is a stagnation of per capita incomes at their present level for the rest of the decade. Only a major increase in commitments of concessional capital within the next two or three years, possibly through the reallocation of capital flows from members of the DAC and from OPEC as well—commitments of a type permitting quick disbursements—can prevent this from happening.

The burden of supplying additional capital would fall heavily on the OECD countries, since they account for a preponderant share of world income and wealth. This would by no means represent an excessive burden, for it would imply no more than maintaining net capital flows to the developing countries at their present relationship to GNP, instead of permitting this ratio to decline.

CHAPTER SIX

The International Financial System

JOHN WILLIAMSON
University of Warwick

THE OIL PRICE INCREASES of late 1973 have brought about major changes in international payments positions and provoked much public discussion and disagreement about the problems that these changes will pose for the international economy. Accordingly, this chapter is concerned with such questions as these: How great is the impact of the oil price increase? Are the problems that it has created really intractable? How has it affected prospects for the industrially advanced countries? for the developing countries? How has international economic policy changed in response? What further responses are in order if the international financial system is to adjust successfully to a permanent increase in the price of oil?

The Impact

As a result of higher oil prices, the oil revenue of the eleven major oil exporters that dominate the Organization of Petroleum Exporting Countries (OPEC) has been estimated by the International Monetary Fund (IMF) and the Organisation for Economic Co-operation and Develop-

THE AUTHOR IS INDEBTED to the participants who discussed the paper in the conference held at the Brookings Institution in November 1974, notably Andrew D. Crockett and Rimmer de Vries, for many helpful comments on an earlier draft. The final version was completed in January 1975.

ment (OECD) to have increased in 1974 by some $70 billion to $75 billion.[1]

It can be expected that all the members of OPEC will tend to expand their imports in response to their increased revenue, but the size and speed of this response will vary greatly. At one extreme are Indonesia and Nigeria, where per capita income is low and development possibilities are enormous. There is a strong presumption that imports of these countries will increase with sufficient rapidity to prevent large and prolonged accumulation of reserves. Although its per capita income is higher, Algeria belongs in this group also; its oil revenues are comparatively modest, while its development expenditures are very large. At the other extreme lie Kuwait, Libya, Qatar, Saudi Arabia, and the United Arab Emirates, where per capita income is high and absorptive capacity is low in relation to the size of oil revenue. Although their imports can be expected to expand rapidly in percentage terms, there is no prospect that these countries will cease to run large current-account surpluses unless and until the production cutbacks designed to maintain price levels become quite large. In between lie Iran, Iraq, and Venezuela. These are middle-income countries with substantial possibilities for development, but their oil revenues are so large that it will be some years before their current-account surpluses can be expected to disappear. In sum, the OECD has assumed that in 1974 OPEC imports will increase some 25 percent more (in volume) as a result of the increase in oil revenue than would otherwise have been the case.[2]

Both the IMF and the OECD have published figures for expected current-account positions for 1974. These are shown in the last two columns of Table 6-1 together with IMF statistics for actual outcomes in the previous two years. By historical standards the reversal of the U.S. deficit in 1973 marked a sudden and sharp change in current-account positions, but as the table demonstrates, the changes between 1972 and 1973 were modest compared to those expected between 1973 and 1974. In fact, it is doubtful whether there have ever before been such abrupt and dramatic changes in current-account positions as those which have resulted from the oil price increase.

The basic assumption made in this chapter is that the present price of oil will be maintained in real terms. Early estimates of future oil revenues

1. International Monetary Fund, *Annual Report 1974;* Organisation for Economic Co-operation and Development, OECD, *Economic Outlook,* 15 (July 1974).
2. OECD, *Economic Outlook,* 15 (July 1974), p. 39.

Table 6-1. *Current Balances*
Billions of dollars

Country or group of countries	1972	1973	1974	1974
United States	−7.1	3.3		−3.3
European Community	9.3	5.9		−15.8
France	1.0	0.5		−7.5
Germany	2.8	6.6	−28	9.0
Italy	2.4	−1.3		−8.3
United Kingdom	0.7	−2.3		−9.0
Japan	7.0	0.2		−4.8
Other industrial countries	1.6	2.1		
OPEC	2.0	4.8	65	
Other developing countries	−8.6	−7.9	−20	
Total (errors and asymmetries)	4.2	8.4	17	

Sources: Columns 1–3: International Monetary Fund, *Annual Report 1974*, Tables 8–10, with supporting detail from IMF. Column 4: OECD *Economic Outlook*, 16 (December 1974), Table 22; these figures are not entirely consistent with the IMF's current balance measure, so the change from 1973 to 1974 should be regarded as only illustrative of orders of magnitude.

were sometimes made by combining the new price level with the volume projections made before the oil price increase. This procedure is clearly erroneous, since it makes no allowance for substitution in production or consumption. It is now generally believed that with present price levels the peak OPEC current-account surplus will not exceed in real terms the $65 billion surplus estimated for 1974 and that after the peak has been reached the surplus will tend to decline as a joint result of economies in energy consumption, increased output of alternative forms of energy, increased oil production in countries that are not members of OPEC, and the rise in imports of the members of OPEC.[3]

The Problems

National Bankruptcy

It has been suggested periodically that the current level of oil prices is so high as to threaten some of the oil-importing countries with "national bankruptcy." This phrase may be interpreted in two senses, the first of

3. Ibid., pp. 94–96; W. B. Dale, "Incremental Oil Deficits and Their Consequences," IMF *Survey*, 9 December 1974, p. 384; *Energy Prospects to 1985* (Paris: OECD, 1975), p. 12. This view is also consistent with the analysis of the international oil market presented in Chapter 7.

which is equivalent to the notion of insolvency. This situation would arise if an oil-importing country were only able to finance the oil imports required for the maintenance of a full-capacity level of output by borrowing and were only able to service that borrowing by further borrowing, so that it gradually surrendered an ever-increasing share of its assets to foreign control. If such a situation should develop, the country's net worth would tend in the course of time to approach zero, in that the whole of its capital stock would ultimately be owned by foreigners, via the direct sale of assets, or indirectly, through the contraction of an equivalent amount of debt.

In order to examine the probability of such a situation's developing, it is appropriate to compare the oil deficit—defined as the deficits in the current accounts of the oil-importing countries with the members of OPEC—with the rate of increase in the capital stock of the oil importers. Such a comparison is of interest for both the aggregate of all oil importers and for individual oil-importing countries, but the aggregate comparison is the crucial one, since the aggregate oil deficit can be redistributed between the oil importers by conventional adjustment policies. Net capital formation of the OECD countries in 1973 was upwards of $370 billion,[4] which would suggest that total net capital formation of all oil-importing countries was of the order of $450 billion in 1973 or $500 billion in 1974. This implies that in 1974 the OPEC current-account surplus of some $65 billion was sufficient to acquire title to about 13 percent of the *increase* in the rest of the world's capital stock.

It might, however, be countered that it is not principally the surplus in a single year that constitutes a problem, but the fact that over a number of years the stock of assets owned by OPEC will accumulate. The question thus posed is what proportion of the total stock of nonresidential business capital of the oil importers might ultimately be owned (directly or indirectly) by the members of OPEC. This issue may usefully be explored in terms of the elementary algebra of compound growth.

Let

K_0 = oil importers' capital stock in year 0

I_t = oil importers' net capital formation in year t

4. International Monetary Fund, *International Financial Statistics,* Vol. 28 (March 1975); the figure is a minimum estimate, since lack of 1973 data led to the use of 1972 figures for eight countries, including France and the United Kingdom. Because for Switzerland a figure for depreciation in 1973 was unavailable, it was assumed to be equal to 42 percent of gross capital formation, the same percentage as in 1969, the last year for which depreciation was reported.

g = growth rate of oil importers' net capital formation

i = real interest rate

p_t = OPEC current surplus as a proportion of oil importers' net capital formation in year t

W_n = foreign assets of OPEC in year n

Then

$$I_t = (1 + g)^t I_0 \text{ and } K_n = K_0 + I_0 \sum_{t=0}^{n} (1 + g)^t.$$

Also,

$$W_n = p_0 I_0 (1 + i)^n + p_1 I_1 (1 + i)^{n-1} + \cdots + p_t I_t (1 + i)^{(n-t)} + \cdots + p_n I_n$$

$$= \sum_{t=0}^{n} p_t I_0 (1 + g)^t (1 + i)^{n-t} \quad (\text{assuming } W_{-1} = 0).$$

The general formula for the proportion of the oil importers' capital stock owned by OPEC is therefore

(1)
$$W_n / K_n = \frac{\sum_{t=0}^{n} p_t I_0 (1 + g)^t (1 + i)^{n-t}}{K_0 + I_0 \sum_{t=0}^{n} (1 + g)^t}.$$

If the OPEC current surplus were maintained at a constant fraction of the oil importers' net capital formation (that is, $p_t = p_0$), then W_n / K_n would grow without limit provided that $i > 0$. It has already been argued, however, that the prospect is for a *reduction* in the real size of the OPEC surplus. Consider the case where the real value of the OPEC surplus is, contrary to this expectation, maintained constant over time ($p_t I_t = p_0 I_0$). Equation (1) then becomes

$$W_n / K_n = \frac{\sum_{t=0}^{n} p_0 I_0 (1 + i)^t}{K_0 + I_0 \sum_{t=0}^{n} (1 + g)^t} < p_0 \sum_{t=0}^{n} \left(\frac{1 + i}{1 + g}\right)^t < p_0 \text{ provided } i < g.$$

Provided that the real rate of growth of the world economy and hence of net investment exceeds the real interest rate, the OPEC share in the wealth of the oil importers would build up to a maximum figure below 13 percent and then decline. If one assumes that the real growth rate of the oil importers' net investment is 5 percent (about equal to that experienced in the past decade), that the real rate of interest is 3 percent (the historical

Table 6-2. *A Comparison of Capital Accumulation and Oil Deficits*
Billions of dollars

Country	Net capital formation, 1972	Increased deficit owing to oil price increase, 1974
Denmark	2.7	0.8
Italy	13	5.2
Japan	57	12
Korea	1.3	0.9
Philippines	0.6	0.5
United Kingdom	14	5.2
Uruguay	0.14	0.09

Sources: Column 1: Derived from International Monetary Fund, *International Financial Statistics*, Vol. 27 (August 1974). Column 2: OECD, *Economic Outlook*, 15 (July 1974), pp. 47–49; Republic of Korea, *Monthly Foreign Trade Statistics*, November 1973; *Foreign Trade Statistics of the Philippines*, 1973; and an estimate of the oil price increase as one of 300 percent. (Depreciation in Uruguay was assumed to have been maintained at 35 percent of gross investment, as in 1970.)

norm), and that the present capital stock of the oil importers is $3 trillion, the OPEC share of the oil importers' wealth would build up to a maximum figure of just over 9 percent in eighteen years and then decline.[5] The proportion could of course go somewhat higher if the West should sell off its capital stock at the knockdown prices reflected in late 1974 stock-market valuations (although OPEC purchasing on a large scale would soon reverse these distress prices).

In the aggregate, therefore, the oil importers face the prospect of losing title to a proportion of the *increment* in their capital stock, rather than to a progressive absolute reduction in net worth. They will of course have to make the psychological adjustment to a situation of substantial foreign indebtedness, but this is nothing novel in modern economic history. The possibility remains that there are some individual countries for which the picture of growing impoverishment is accurate. A comparison between net capital formation in 1972 and the initial impact of the oil price increase for a number of particularly hard-hit countries is therefore made in Table 6-2. (This comparison is biased in the pessimistic direction, in that it compares investment in 1972 dollars with an oil deficit measured in 1974 dollars.) It will be observed that none of the industrialized countries faces the pros-

5. This figure may be compared with the 48 percent of the Argentine capital stock that has been estimated to have been foreign-owned in 1914. See Carlos F. Díaz Alejandro, *Essays on the Economic History of the Argentine Republic* (Yale University Press, 1970), p. 30.

pect of having to borrow a sum approaching the size of its new investment but that some of the developing countries are in a less happy situation. Unless they reduce consumption and increase savings, either to step up domestic investment or to reduce their current deficits, they face the prospect of a severe slowdown in the rate at which they have been accumulating wealth.

Rough as these comparisons are, they will suffice to demonstrate that there is no prospect that oil deficits alone will drive some of the oil-importing industrial countries to national bankruptcy, in the sense of ever-increasing external debt in relation to debt-servicing capacity, and this is so even without a redistribution of the oil deficit through adjustment policies. There is, however, a second, if less dramatic, sense in which the term "national bankruptcy" might be interpreted, which relates to the concept of illiquidity rather than that of insolvency: a situation might develop in which some oil-importing countries would be unable to meet their contractual obligations to the outside world, or, more generally, would be faced with an acute liquidity problem. This is a problem that cannot arise for the oil importers in the aggregate, since all the important methods by which the members of OPEC receive payment involve their acquisition of some form of claim on one or more oil-importing countries. It could, however, arise for individual countries, if the process of financial intermediation needed to enable OPEC members to acquire title to an average 10 to 20 percent of the new investment in the oil-importing countries were to operate in a way which denied certain countries adequate access to credit.

Financing

The question at issue is whether the financial mechanism will be capable of recycling the oil revenues in such a way as to finance a "desirable" set of current-account deficits on the part of the oil importers. Failure to achieve an appropriate distribution would involve at best a misallocation of resources and at worst a major slump or acute liquidity shortages.

Oil payments are made almost entirely by countries drawing on their foreign-exchange reserves and transferring foreign exchange to the members of OPEC: the use of primary reserve assets—gold and special drawing rights (SDRs)—is minimal and seems likely to remain so, in view of the financial sophistication of the oil exporters and the low interest rate on the SDR. Some countries, pricipally certain developing countries with close ties to certain OPEC members and some industrialized countries which have concluded barter arrangements, have lines of credit which avoid or reduce the need for transfers of foreign exchange, but these are not large enough to influence significantly the conclusions reached by assuming that all oil payments are made in terms of foreign exchange.

Until now the principal recycling mechanism has been the Eurocurrency market. If an oil importer draws on reserves held in a reserve center, while the OPEC member places its new reserves in the Euro market, the Euro bank acquires additional funds which can be lent to the oil importers. If public- or private-sector borrowers in each oil-importing country borrow from the Euro banks funds equal to the amount of each country's oil deficit and convert these funds into domestic currency, and if the central banks then place their additional reserves in the same way that they placed their previous reserves—that is, in the reserve center—global reserves will rise by the size of the aggregate oil deficit and oil importers' reserves will remain unchanged. No liquidity problem will arise so long as this process continues.

Since the increased oil payments started, global reserves have risen by SDR 19.8 billion (from SDR 152.5 billion in December 1973 to SDR 171.3 billion in August 1974) while the reserves of ten principal oil exporters have risen by SDR 18.1 billion.[6] The outcome so far is therefore very close to that predicted by the basic model of the recycling process. There are a large number of other ways in which the intermediation process could work, however, and some of these lead to different conclusions.

1. The oil exporters may invest directly in the oil-importing countries, rather than in the Eurocurrency markets. Failure to use the Euro market as an intermediary changes nothing essential, although if the assets thus acquired do not have the status of reserve assets, global reserves will not expand. The important point, however, is that the oil importers' reserves will still remain unchanged, both in the aggregate and, to the extent that capital inflows match current deficits, those of individual countries.

2. The oil exporters may invest their reserves in the issuing country rather than in the Euro market. If the reserve center were surrounded by an effective exchange-control fence which prevented capital outflows, the result would be that the reserves of the oil-importing countries (other than the reserve center) would decline by a sum equal to their oil deficits, and

6. IMF, *International Financial Statistics*, Vol. 27 (November 1974), p. 18.

world reserves would remain constant. At the other extreme, if there were perfect capital mobility, the decision by the oil exporters as to where to place their funds would make no difference: oil importers unable to borrow from the Euro market would simply borrow directly from the reserve center instead. In between lies reality: because of imperfections in capital mobility, and because the placement of reserves in the reserve center provokes monetary contraction by the monetary authorities insofar as they determine their actions in the light of domestic conditions, the result would be a capital outflow which would *partially* replenish the reserves of the other oil importers.

3. The oil importers may draw down their reserves held in the Euro markets rather than those held in the issuing country. If the oil exporters placed their reserves in the Euro markets, there would be no change in the funds available in the Euro banks for lending, and hence there would be no change in global reserves with consequent fall in the reserves of the oil importers. If the oil exporters placed their funds in the issuing country, the lending power of the Euro market and the level of world reserves would decline, and the reserves of the oil importers would decline by twice the size of their oil deficits. Once again, however, such declines in the reserves of the oil importers would prompt them to borrow more from the reserve center, with a consequent mitigation of the decline in reserves.

4. Some oil importers, namely, the reserve centers, may pay the oil exporters by issuing additional reserve liabilities rather than by transferring reserve assets. If the oil exporters should hold these reserves in the issuing countries, the result would be the same as in the basic model or in variant (1): the reserves of the oil importers would remain unchanged. If the members of OPEC placed their reserves in the Euro markets, the lending potential of the Euro markets would rise and the reserves of other oil importers would rise also. This effect would tend to be mitigated by a capital flow toward the reserve centers.

5. The borrowers in the oil-importing countries may not convert the entire proceeds into domestic currency. If they should continue to hold some of the borrowed funds in the Euro markets, or if they should use them to pay those who deposit their funds in the Euro markets, this would give rise to a Eurocurrency multiplier in excess of unity. Reserves would not, however, increase as a result, since borrowing in the Euro market leads to an increase in reserves only when the proceeds are converted into domestic currency. The principal reason that the Eurodollar multiplier

has in the past slightly exceeded unity is that a number of the central banks redeposited reserves in the market, an activity which does indeed lead to an increase in the level of reserves. The implication is that any decrease (or increase) in the reserves of the oil importers as a result of the preceding considerations will be somewhat magnified.

The above analysis points to a reasonably clear conclusion. Where the members of OPEC place their increased revenues and whence the oil-importing countries draw their reserves will have some impact on the ease with which the oil importers can finance their deficits. However, so long as there is reasonable mobility of capital among the oil importers, including the reserve centers, there is no real danger that a creditworthy country prepared to borrow will face the prospect of an inability to do so as the consequence of a global liquidity shortage.

There remain a number of crucial questions about the intermediation process. The first relates to the ability of the Euro market to continue playing the dominant intermediary role. It is feared that the market may collapse, or, less dramatically, that it may be unable to continue expanding at an adequate rate. The danger of collapse would arise only in the event that the central banks should be unwilling to step in to stem any run that might develop. It is true that the central banks have been hesitant to give assistance to some banks whose unwise activities have got them into trouble, thereby creating difficulties for the smaller Euro banks; but none of the problems so far has threatened to precipitate a cumulative run on the 1931 model, and it is difficult to believe that such a threat would be allowed to develop. It would, nevertheless, undoubtedly be reassuring to know that the central banks had agreed as to which one of them should take responsibility in the ambiguous situations created by multinational banking (where a bank located in country A may lend the currency of country B deposited by residents of country C to those of country D). A somewhat more real danger is that the market may run into constraints on its expansion. It has, for example, been asserted that the size of the capital base of the Euro banks will soon start discouraging them from accepting new deposits. Obviously such constraints are not absolute. A widening in the spread between borrowing and lending rates would raise the profitability of Euro banking, thereby helping to attract the equity capital necessary to finance a further expansion of the market. It is equally true that a lowering of the borrowing rate would intensify the pressure on the members of OPEC to lend directly to the oil importers rather than use the Euro market as an intermediary, something which seems to have occurred in late 1974; as noted earlier, this can provide an equally satisfactory method of recycling.

Another difficulty which has been asserted is a mismatch in maturity preferences. Specifically, the oil exporters like holding short-term assets, while the oil importers would prefer to borrow long. This situation is a classic example of one which the price mechanism can be expected if not to resolve at least to minimize: short-term rates will fall and long-term rates will rise, until the lenders or the borrowers—or both—are induced to modify their preferences sufficiently to overcome the mismatch in preferences.[7] The process will no doubt be facilitated by the continued enterprise of the international bankers, whose invention of rollover loans and floating rates has gone a long way toward simultaneously satisfying the desire of borrowers for an assurance of long-term credit with that of lenders for an avoidance of long-term commitments unaccompanied by indexing.

There seems therefore no compelling reason for fearing a failure of the recycling process which would lead to a global liquidity shortage. The real danger is that certain individual countries may cease to be creditworthy and therefore unable to borrow despite the continuing adequacy of global liquidity. The central question is, therefore, What is it that determines creditworthiness? A traditional answer, based on the real-bills doctrine, might be that a country can be judged creditworthy only so long as there is a clear prospect of its continuing to move back to a position of surplus, so that it will be able to repay its debts. If this criterion were applied it would be certain that a large number of countries would have to be judged not worthy of credit, since the oil importers in the aggregate are likely to continue to have substantial current-accounts deficits for an indefinite period.

7. Another fear which in present circumstances does not seem particularly well founded is that such a decline in short-term interest rates will provoke members of OPEC into restricting oil production. Such restrictions would be rational only if oil in the ground were expected to be a more profitable investment than assets on paper. In view of the near certainty that the relative price of oil will decline in the future unless constraints upon output are progressively tightened, the present value of oil in the ground is markedly less than its current price, and it is therefore less than the present value of the financial returns that can be expected from present production, unless the real rate of interest on financial assets should prove to be significantly negative.

But such a criterion would be no more rational than the real-bills doctrine on which it is based. A far more appropriate criterion is a clear prospect that indebtedness will not grow faster than debt-servicing capacity—which is to say, that a country remains solvent. This is a criterion which, as argued in the preceding section, individual countries can certainly hope to meet, provided that the current-account deficit is appropriately distributed.

This is not to argue that there is no legitimate role for "lender of last resort" facilities provided under official auspices. In the first place, the superimposition of the oil deficit means that countries with large nonoil deficits must be markedly less creditworthy than would otherwise have been the case. Even if the correct criterion of creditworthiness is applied, therefore, there are likely to be countries which periodically have a need for credit from official sources. In the second place, there is no certainty that the banks will not insist upon an inappropriately rigorous standard of creditworthiness. Third, confidence is also to some extent dependent on psychological attitudes that may not always be entirely rational. Adequate official recycling facilities are therefore an important safeguard for the stability of the system. But at the same time, official recycling facilities too easily made use of could be dangerous. The reason is that decreasing the pressure upon countries to achieve appropriate current-account targets (which is the prime requirement for the maintenance of creditworthiness) might lead to a concentration of deficits in the countries with the weakest political determination to avoid deficits, thus also enabling countries with xenophobic objections to foreign indebtedness to evade their fair share of the debt burden. In sum, official recycling should not be conducted on particularly favorable terms and a program offering a clear prospect of pruning back the size of current deficits to internationally agreed-upon target levels should be a condition of its use.

Global Demand Management

It has frequently been pointed out that the increase in oil prices will have effects analogous to those of an increased excise tax, to the extent that oil revenues are not spent by their OPEC recipients on additional imports. These effects are to increase prices (at least in the short run, when the cost-inflationary effects are dominant) and to decrease real demand.

The OECD has estimated that on average the direct effect of the oil price increase will be to increase prices generally, as measured by gross domestic product (GDP) deflators, by 1.5 percent.[8] This estimate needs to be increased to allow for the effect of the oil price increase in pulling up the prices of other fuels in sympathy and for the repercussions resulting from the wage-price spiral. The total impact of the oil price increase on the general price level is therefore likely to be quite sizable by historical standards, but it can hardly be considered the principal cause of the present global inflation.

The direct deflationary impact on demand stems from the increased saving being undertaken by the oil exporters and is therefore related to the increase of some $60 billion in the OPEC current-account surplus. This direct impact is increased by the usual multiplier effect, which is normally calculated on the assumption that monetary policy is directed to the maintenance of a constant interest rate, and—under any other assumption about monetary policy—is further modified by the monetary repercussions of a tax change. These monetary repercussions depend (in the domestic context) on the use to which the increased tax revenue is put: If it is used to reduce debt held by the public, interest rates will fall and stimulate a rise in spending that will partially offset the initial decline in demand, whereas if it is used to reduce the money supply (for example, by paying off debt held by the central bank), interest rates will rise, thereby intensifying the initial decline in demand. It is therefore relevant to consider whether in the international case the increased oil revenue will lead to a monetary contraction. The analysis of the previous section implies that while some decline in reserves (and therefore in domestic money supplies, insofar as oil importers do not sterilize but instead follow the traditional rule of linking changes in the money supply to reserve changes) may occur, it is unlikely to be a major decline. Its impact in any country, moreover, can be offset by action of the central bank. Hence the correct analogy is to an increased excise tax whose proceeds are used to retire debt rather than to reduce the money supply. In the case of the oil surplus, the equivalent to debt reduction is the increase in oil funds seeking investment outlets. Like debt retirement, this will offset to some extent the reduction in demand produced by the fall in real income, even in the absence of deliberate reflationary measures.

There is another and, in the long run, possibly more important difference from the case of an excise tax. Excise-tax increases rarely create large new investment opportunities; the oil price increase, on the other hand,

8. OECD, *Economic Outlook*, 15 (July 1974), p. 30.

increased investment demand, specifically in the energy industries. It is generally considered unlikely, however, that the additional capital spending in the energy industries of the oil-importing countries will match the size of the increased OPEC saving,[9] and there may be a contraction in investment in industries that produce goods complementary to oil (such as automobiles). There is therefore a need, especially in the short run, for reflationary measures to limit the decline in output.[10]

Current-Account Targets

The arguments offered earlier for seeking an agreed-upon and appropriate basis on which to distribute target current-account deficits are reinforced by the danger that in the absence of such an agreement countries might resort to competitive payments policies in a series of national attempts to eradicate deficits that are collectively inconsistent. Given the low short-run substitutability between oil and other products, adjustment policies could not significantly curtail the collective oil deficit except to the extent that output fell, irrespective of whether the actual policies adopted consisted of deflation, depreciation, or import restrictions.

The danger of competitive payments policies was taken seriously by the international financial community. For example, the Rome communiqué of the Committee of the Board of Governors of the International Monetary Fund on Reform of the International Monetary System and Related Issues (the Committee of Twenty) in January 1974 included the initial injunction to "accept the oil deficit" in the following terms:

They recognized that . . . many [nonoil-exporting] countries . . . would have to have large current account deficits. In these difficult circumstances the committee agreed that, in managing their international payments, countries must not adopt policies which would merely aggravate the problems of other countries. Accordingly, they stressed the importance of avoiding competitive depreciation and the escalation of restrictions on trade and payments.[11]

9. T. M. Rybczynski, in "Capital Requirements for Energy Development in the Main Industrial Countries" (Paper presented to the British–North American Association, April 1974), however, estimated that the announced energy investment programs of the OECD countries would substantially exceed the increase in OPEC saving. Recent OECD estimates also seem to suggest that the increase in energy investment may be of the same order of magnitude as the increase in saving; see *Energy Prospects to 1985,* Part II.

10. The net deflationary effect of higher oil prices on the industrial countries proved to be very large, as discussed in Chapters 1–4. Also, see the discussion of energy investment in Chapter 1, pp. 29–33.—Eds.

11. IMF *Survey,* 21 January 1974, pp. 17, 22.

The danger of competitive payments policies has not materialized so far. The French decision to float was initially regarded as a possible example of a predatory policy, but subsequent French action in using reserves to limit depreciation, and in borrowing substantial sums, allayed this fear. The Rome consensus may have contributed to this welcome resolution, but one may doubt whether it is in any considerable measure attributable to international co-operation. The fact is that the competitive payments policies of the 1930s were not provoked by a desire for current surpluses per se, but for the employment-creating effect of a current-account surplus. The powerful influences on politicians come from their internal constituencies, who are concerned about jobs and prices, rather than from their external obligations. The strength of international co-operation is therefore being tested only as the recession develops.

One is therefore confronted with the question as to how a set of current-account objectives should be determined, which is another way of asking the question as to what the injunction to "accept the oil deficit" is supposed to mean. Should each oil-importing country aim for an increase in its target current deficit equal to the increase in its bilateral deficit with the members of OPEC? Or should the oil importers agree merely to accept the oil deficit collectively, rather than individually? In that event, how should the collective deficit be distributed? By reference to capital flows, by which is meant the geographical pattern of investment chosen by the oil exporters?

Doubtless a rule that each country should accept its individual oil deficit provides a useful short-run guide in a situation where *some* rule is needed and negotiation of a complex rule requires time, but as a long-run guide it makes no economic sense at all. In order to maintain the level of employment, the oil-importing countries in the aggregate have to reduce savings and/or increase investment by a sum equal to the increased OPEC savings. Each oil-importing country can be expected to reduce its savings voluntarily by a sum equal to its marginal propensity to save multiplied by the loss in real income attributable to the oil price increase, said loss being its initial individual oil deficit *before* making any allowance for increased OPEC imports. The oil importers in the aggregate must then increase investment or reduce saving further by the difference between the increased saving of the oil exporters and the voluntary reduction in the saving of the oil importers, in order to avoid deflationary pressure. Economic efficiency clearly requires that increased investment should be distributed geographically with reference to the productivity of investment, rather than located disproportionately in those countries which suffered the largest cuts in real income (as would be required if countries were expected to accept their individual oil deficits).[12] To take a concrete example, the criterion of individual acceptance of oil deficits would require Canada to decrease its current deficit, the absurdity of which is self-evident in view of the opportunities for profitable energy investment within Canada that have been created by the oil price increase. Insofar as the deflationary gap has to be closed by an involuntary reduction in saving, the natural principle on which to base its distribution is in proportion to GNP.

The preceding argument concerning the determination of appropriate changes in current-account targets may be expressed algebraically.

Let

ΔR = increase in payments for oil
ΔM = increase in OPEC imports
$D = \Delta R - \Delta M$ = oil deficit
Y = real income
S = savings
I = investment
s = marginal propensity to save
T = target current-account deficit
i = ith oil importer
v = voluntary
f = involuntary ("forced").

Preservation of "full employment" requires that $D - s\Delta R$, the increase in OPEC savings minus the voluntary decrease in the oil importers' savings, be matched by increased investment, ΔI, and involuntary reductions in savings, among the oil importers in the aggregate. In principle the increased investment should be located where the marginal efficiency of investment is highest. In the absence of any specific information about rates of return, it is plausible to suppose that this would be approximately realized if the increase in investment, like the involuntary reduction in saving, were distributed in proportion to GNP; or $\Delta I_i = (Y_i/Y)\Delta I$, just as $\Delta S_i^f = (Y_i/Y)\Delta S^f$. Hence

$$\Delta T_i = \Delta I_i - \Delta S_i^f - \Delta S_i^v$$
$$= s\Delta R_i + (Y_i/Y)[(1 - s)\Delta R - \Delta M]$$

12. This conclusion has been urged by W. M. Corden and Peter Oppenheimer in "Basic Implications of the Rise in Oil Prices," *Moorgate and Wall Street,* Autumn 1974, pp. 23–38.

on substitution. Thus the target current-account deficit should increase by a sum equal to the individual deficit before allowing for increased OPEC imports (ΔR_i), multiplied by the marginal propensity to save, plus a share of the collective deficit proportionate to GNP, after allowing for increased OPEC imports and also subtracting that part of the oil deficit which is to be absorbed at the point of impact.[13]

The implications of this approach for the design of adjustment policies can be more readily appreciated by introducing some additional notation:

$$y_i = Y_i/Y = i\text{th country's share of total oil importers' GNP}$$

$$r_i = \Delta R_i/\Delta R = i\text{th country's share of additional oil bill}$$

$$\rho_i = r_i/y_i = \text{ratio of } i\text{th country's additional oil bill as proportion of GNP to total oil bill as proportion of oil importers' GNP}$$

$$m_i = \Delta M_i/\Delta M = i\text{th country's share of additional exports to OPEC}$$

$$\mu_i = m_i/y_i = \text{ratio of } i\text{th country's additional exports to OPEC as proportion of GNP to total additional exports to OPEC as proportion of oil importers' GNP.}$$

Assuming that a country had an appropriate current balance prior to the oil price increase, the adjustment action needed as a result of the higher oil price can be found by subtracting the oil deficit from the increase in the target current-account deficit. Substitution eventually yields

$$\Delta T_i - D_i = y_i[(1 - \rho_i)(1 - s)\Delta R - (1 - \mu_i)\Delta M].$$

A country with an average share of both R and M (that is, $\rho_i = \mu_i = 1$) would therefore be expected to accept its individual oil deficit ($\Delta T_i = D_i$). A country with above-average dependence on oil imports ($\rho_i > 1$), or with a below-average share in additional exports to OPEC ($\mu_i < 1$), would be expected to take adjustment actions (such as depreciation of its exchange rate) designed to reduce its deficit below its oil deficit ($\Delta T_i < D_i$). Countries in the reverse situation would be expected to adjust—by appreciation, for example—so as to run a nonoil deficit. These countries would of course have to stimulate investment or involuntarily reduce saving by more than the initial deficiency in aggregate demand, while those countries required to run nonoil surpluses would need to restrict reflationary measures to less than those required to offset the initial decline in demand internally.

13. I am indebted to J. Marcus Fleming for suggesting this approach.

Table 6-3. *Approximate Estimates of Appropriate Increases*
in Target Current-Account Deficits for Major Industrial Countries
Billions of dollars

Country	Voluntary decrease in savings	Distribution according to GNP	Total (rounded)
United States	3.4	15.8	19
Germany	1.3	4.1	5
France	1.3	2.9	4
Japan	2.5	5.2	7½
United Kingdom	1.0	2.0	3
Italy	1.0	1.7	2½

Some rough calculations of the increased target current-account deficits for the major oil importers that would be implied by the above approach are shown in Table 6-3. The calculations assume that the marginal propensity to save is 0.2 in all countries, which implies that some $15 billion of the oil deficit is voluntarily absorbed at the point of impact and that OPEC imports have risen by $15 billion; this leaves some $45 billion of the oil deficit to be distributed in proportion to GNP. The aggregate oil importers' GNP is taken to be $4 trillion.

The formulas developed in this section might be questioned by some on the grounds that it is inappropriate for countries to develop specific current-account targets. In this view each country should adopt current-account adjustment actions designed to keep the overall balance of payments in equilibrium—that is, to secure a transfer of real resources to offset any flow of financial capital. However, given the capacity of governments to influence the flow of financial capital—a capacity which was dramatically illustrated during 1974—such an approach is tantamount to allowing a free-for-all in the determination of payments objectives. Moreover, even if one did succeed in preventing governments from adopting measures aimed at manipulation of the capital account, such an approach would create a real danger of building an undesirable element of instability into the system. The reason is that preoccupation of the OPEC investors with security would tend to lead them into investing in the countries with the largest current surpluses and shunning those with large deficits; the former countries would then be enjoined to eliminate their surpluses and replace them with deficits to match the capital inflow, and the deficit countries would be under pressure to do the opposite. As the prescribed adjustment actions took effect, the OPEC investors would tend to reverse the distribution of their investments, thus requiring a reversal of adjustment policies. There is thus no real alternative to attempts by governments to determine on rational grounds where additional investment should take place; calculating the current-account implications of this distribution; and then seeking to ensure that capital flows are such as to finance the desirable current-account imbalances. The real question is whether these calculations will be made collectively, and therefore consistently, or whether the dangers of competitive policies or maldistribution of the deficit, or both, will be realized.

Disequilibrating Capital Movements

Since the members of OPEC are accumulating large portions of their additional portfolios in highly liquid assets, they will be in a position to switch increasingly massive sums from one currency to another. There are those who see in this ability real dangers—either of the oil exporters' making speculative profits, or of their using their financial power to disrupt the economy of the OECD area.

Since the adjustable peg has now been abandoned, the opportunities for unlimited risk-free profits from speculative switching have disappeared. Some economists would argue that, under floating, there is no scope for profitable destabilizing speculation.[14] I have elsewhere disputed the proposition that destabilizing speculation cannot in principle be profitable,[15] and the model that I constructed would seem more likely to apply where the potential speculators have some monopoly power.[16] Nevertheless, it is difficult to believe that the danger is real. If the oil exporters knew enough about the relevant lags and elasticities in the trade flows to exploit the possibilities of making speculative profits, it seems rather unlikely that the central banks of the industrialized countries would be ignorant of such knowledge and thus unable to undertake a highly profitable strategy of stabilizing counterspeculation.

14. Cf. Milton Friedman, "The Case for Flexible Exchange Rates," in his *Essays in Positive Economics* (University of Chicago Press, 1953), pp. 157–203.
15. John Williamson, "Another Case of Profitable Destabilizing Speculation," *Journal of International Economics,* Vol. 3 (February 1973), pp. 77–83.
16. L. D. Price and G. E. Wood, "Another Case of Profitable Destabilizing Speculation: A Note," *Journal of International Economics,* Vol. 4 (May 1974), pp. 217–20.

The other possibility is that some oil producers might adopt a politically motivated campaign of economic disruption, even if this should be likely to cost them something. The past record of most OPEC members as rather cautious investors does not make such actions seem particularly likely, and their growing stake in the Western economies—and growing need to avoid provoking thoughts of nationalization—would seem likely to make such disruption increasingly unlikely. In addition there remains the residual safeguard provided by the unlimited collective ability of the Western central banks to counterspeculate. There seems little reason, therefore, for treating the accumulation of liquid assets in OPEC hands as any different from the general growth of mobile liquid assets.

Income Distribution

All the problems caused by the oil price increase that have been asserted as posing major dangers for the functioning of the international economy have been considered in the preceding sections. The discussion provides no basis for the belief that these problems are unmanageable; on the contrary, while international co-operation to secure a rational distribution of the deficit is indeed essential, this is a need that is very much amenable to rational economic management.

If this conclusion is correct, it might suggest one's puzzling over the question of why the oil price increase has provoked such apocalyptic expressions of doom. The puzzle is not, however, a profound one, at least to those who have some sense for political economy and who do not regard questions of income distribution as extrascientific and therefore unworthy of recognition. The central fact about the oil price increase is simply that it transferred some $70 billion to $75 billion per annum from the consumers to the producers of oil, which transfer is very much a "problem" to the first of these two groups. No one likes to lose income—especially, perhaps, when it is redistributed to someone else. To some economists it makes matters considerably worse that the redistribution was effected by the exercise of monopolistic power rather than by the play of market forces. Moral disapproval of the actions responsible for the income transfer, however, scarcely justifies a search for reasons to condemn the transfer as leading to technically unmanageable problems.

Given this diagnosis of the central issue, it may seem surprising that the rich countries, which can afford the income loss involved with relative

Table 6-4. *Effects of the Oil Price Increase on the International Distribution of Income*

Country	1971 per capita GNP in dollars	Approximate 1974 per capita GNP in dollars	1974 Income transfer in billions of dollars
OPEC			
Kuwait	3,860	12,000	+8
United Arab Emirates	3,150	19,000	+3
Qatar	2,370	17,000	+1
Libya	1,450	6,000	+5
Venezuela	1,060	2,000	+9
Saudi Arabia	540	3,500	+18
Iran	450	1,300	+14
Iraq	370	1,100	+4
Algeria	360	750	+2
Nigeria	140	250	+5
Indonesia	80	125	+3
OECD	3,160	4,700	−63
United States	5,160	6,600	−17
France	3,360	5,200	−6½
Germany	3,210	6,400	−6½
United Kingdom	2,430	3,300	−5¼
Japan	2,130	4,700	−12½
Italy	1,860	2,700	−5¼
Non-OPEC developing countries	60–1,230	n.a.	−10

Sources: *World Bank Atlas*, 1973; OECD *Economic Outlook*, 16 (December 1974), Table 19, p. 54; *BP Statistical Review of the World Oil Industry*, 1973; Edward R. Fried, "Financial Implications," in Joseph A. Yager, Eleanor B. Steinberg, and others, *Energy and U.S. Foreign Policy* (Ballinger Publishing Company, 1974), Table 14.2, p. 283; Banque de Bruxelles, *Bulletin Financier*, October 1974.

ease, should have exhibited a far more hostile reaction to the oil price increase than have some of the developing countries. An explanation of this paradox might be sought in the markedly differing views of the two groups of countries regarding the legitimacy of the way in which income was distributed before the oil price increase.

Be that as it may, it is clearly relevant to seek to establish some of the basic facts about the effect of the oil price increase on the international distribution of income. Table 6-4 represents an attempt to assemble such information. It will be observed that it is only the Gulf sheikhdoms that have enjoyed gains in income large enough to raise their income markedly above the general OECD level, and only some 15 percent of the total transfer of income has accrued to them. At the other extreme, some 40 percent of the increased oil revenue has accrued to countries which are still unambiguously poor by OECD standards. Since over 85 percent of the total loss of income has been sustained by the OECD countries, it is far from obvious that the overall effect on income distribution can be considered perverse. The worst that can be said is that the effect is capricious; much the same might be said of the distributional consequences of domestic cost inflation.

The Responses of International Policy

The preceding analysis of the problems posed to the international economy by the oil price increase led to the conclusion that the fundamental issues concern the "real" problems of waste of resources and distribution of income, rather than the financial questions on which so much attention has been lavished. This is not to say that the financial issues are unimportant, but rather that—given a reasonably appropriate distribution of the oil deficit—they should be capable of satisfactory resolution.

The remainder of this chapter provides a brief survey of the ways in which international economic policies have already responded to the oil price increase and some suggestions as to the additional responses that may be looked for in the future.

Financial Intermediation

There were three major policy initiatives addressed to the problem of financial intermediation during the first half of 1974. The first was the action of the United States in abolishing controls over capital outflows. This had the effect of greatly reducing the importance of decisions by members of OPEC as to where to place their funds, since it meant that a failure by some oil-importing country to attract OPEC investments, directly or through the Euro markets, could be compensated for by increased borrowing from the United States. The second initiative was the creation of the IMF "oil facility" following the final meeting of the Committee of Twenty in June. This facility borrows funds directly from the oil exporters and lends them to those oil importers with a balance-of-payments need, up to a maximum determined by a formula that takes into account the increased cost of oil imports, the strength of the country's reserve position, and the country's Fund quota.[17] This facility is of particular usefulness to those countries faced with problems of creditworthiness. The third initiative was the active solicitation of OPEC investments by the International Bank for Reconstruction and Development (IBRD, the World Bank). This can be expected to benefit principally the higher-income developing countries.

The severity of the problems facing certain countries on account of diminished creditworthiness suggests that there will be a need for additional mechanisms to supplement the IMF oil facility. There would seem to be two general patterns which such new mechanisms might follow. One would involve the oil exporters' taking equity-type investments in the countries involved; this would give the investors both protection against unforeseeable changes in the rate of inflation and an expectation of good returns in the long run without adding correspondingly to the medium-term indebtedness of the host countries and thereby their problem of creditworthiness. Individual investments—whether in the form of direct investments or the purchase of existing equities—would of course be quite risky, but the oil exporters' portfolios are sufficiently large to enable them to cope with problems of risk by diversification. The second pattern would involve further use of the international financial institutions. The risk of default would thereby be reduced so far as the OPEC lender is concerned, and perhaps—given the reluctance of countries to incur the disapproval of the whole international community—the risk of default by the borrower would also be reduced. There are a wide variety of organizational forms that could be utilized—the sale of additional IBRD bonds, the issue of additional SDRs (although these would have to bear a sufficiently high rate of interest to make them appealing to the oil exporters as an investment medium), or the expansion of the oil facility as agreed upon by the IMF Interim Committee in January 1975.

An important question that would be raised by extensive use of the international institutions as intermediaries concerns the monetary unit to be used in denominating liabilities. The IBRD still uses national currencies —frequently the currency of the lending country—for this purpose. Given the absence of any balance-of-payments constraint on the part of many OPEC members, this amounts to giving the lending country the right to write up the value of its assets—and other countries' debts—unilaterally. Irrespective of the confidence one may have that any particular country

17. IMF *Survey*, 16 September 1974, pp. 299–301.

would not take advantage of such an opportunity, it seems highly doubtful whether the creation of this kind of temptation is either wise or morally justifiable. The Fund's definition of the SDR in terms of a basket of currencies has the effect of freeing the value of the SDR from the unilateral control of any country, thereby giving the SDR one important feature of a satisfactory international, intertemporal unit of account. However, the SDR still suffers from the disadvantage that its purchasing power is subject to erosion from inflation in the countries whose currencies compose the basket. A known rate of inflation could be (though the present rate of inflation has not been) offset by an appropriate rate of interest, but no fixed interest rate can eliminate the risk—to both lenders and borrowers—stemming from unanticipated changes in the rate of inflation. Insofar as both lenders and borrowers are averse to risk, both could benefit by the adoption of a unit of account defined in real terms—that is to say, by indexing. Apart from objections to indexing on the grounds of its supposed inflationary effects, however, there is also the statistical problem of finding a satisfactory general index of the price of goods traded internationally which might present an obstacle to rapid implementation of any proposal involving indexing.

The preceding analysis has not suggested that there is a compelling need for the creation of any new intermediation agency such as the $25 billion OECD safety net or solidarity fund originally suggested by Henry Kissinger on 15 November 1974. One reason for doubting the value of such an agency is that most of the countries likely to suffer constraints on their borrowing ability by virtue of limited creditworthiness are developing countries and would therefore be excluded from the scheme. A second reason stems from the fact that a country having extensive private foreign indebtedness would not be made significantly more creditworthy by such a scheme if a part of its borrowing were from intergovernmental rather than private sources; all of the developed countries—unlike many of the developing countries—are in that position and would therefore still be subject to the disciplines of the private market. But, provided it did not undermine discipline, a lender-of-last-resort facility would be marginally useful in providing a safeguard against the danger of private banks' limiting new business in one way or another—because of a failure to modify their standards of creditworthiness, for example. One may, however, regret that this safety net would apparently exclude qualified developing countries from its scope, although the ill effects of this may in practice be minimal if the IMF oil facility directs most of its lending to the excluded countries.

Adjustment Policy

As noted earlier, the danger that the oil price increase would stimulate competitive adjustment policies has been taken seriously by the international community. The injunction of the Committee of Twenty in January 1974 to countries to "accept the oil deficit" was followed in May by the pledge by the members of the OECD that they would refrain from introducing or intensifying import restrictions for a one-year period and in June by the agreement of the Committee of Twenty that the IMF would sponsor a voluntary "trade pledge," by which subscribing members would oblige themselves to seek the Fund's approval before intensifying current-account restrictions.

Despite these international agreements, the year 1974 was not entirely free of restrictive policies. The two most important cases were Italy's imposition of a 50 percent deposit on imports and Denmark's raising of indirect taxes on a large number of items with a high import content. Many smaller countries have also taken restrictive actions. It must be observed, however, that all the countries involved have had large nonoil deficits which clearly required remedial measures of one kind or another. One may be critical of the particular measures adopted, but they cannot be seen as part of a competitive scramble to offload the oil deficit onto other countries. Nevertheless, the fact that the danger has not yet materialized is no reason for not trying to strengthen further the defenses against its realization, especially if the view that the incentive for competitive policies would be greatly intensified by recession are accepted.

The obvious measure to secure this strengthening is explicit agreement on a set of consistent current-account targets. It has been argued above that such an agreement is also of prime importance in the attempt to secure a wide distribution of the oil deficit so as to safeguard the creditworthiness of individual countries. The principles which should underlie such an agreement were discussed earlier.

International Monetary Reform

When the Committee of Twenty decided, in January 1974, to abandon the attempt to write a new monetary constitution for the world, the principal justification offered was the uncertainties created by the oil developments. In fact, however, the oil situation was an excuse for, and not a cause of, the abandonment of international monetary reform. The reform effort had failed before the October war set in motion the events that revolutionized the oil market. Why it failed is an interesting question that cannot be pursued here. The extent of its failure can be gauged from the facts that the only aspects of reform agreed upon—notably the envisaged exchange-rate regime—were patently unworkable, while a wide range of issues of secondary importance, such as the use of objective indicators of the need for adjustment, had provoked doctrinal conflicts that showed no signs of being bridged in the *First Outline of Reform* published in September 1973.

It is difficult to identify any analytical reason for supposing that the organization of the international monetary system should depend on the price of a particular commodity, even of the most important single commodity entering international trade. Any satisfactory system would need to make provision for the orderly adjustment of current-account positions in response to disturbances; for the channeling of flows of financial capital to areas where investment opportunities exceed local savings; for the elimination of the incentives for disequilibrating capital movements; and for the investment of substantial sums by countries that run prolonged current-account surpluses. These needs were emphasized, but not created, by the oil developments. If some of the Committee of Twenty's negotiators from the industrialized countries failed to appreciate these needs prior to the oil developments, it is they, not OPEC, who should bear the onus for the failure of the Committee of Twenty.

Aid

Since the central fact about the oil price increase is that it has caused the largest sudden international redistribution of income in history, it is only to be expected that the most critical problem the action has raised is the loss of real income sustained by those least able to afford such a loss. These are, of course, the developing countries, and, more specifically, the poorest among them. The nonoil-exporting developing countries have sustained a loss of some $10 billion per annum, of which approximately $2 billion is accounted for by the poorest among them. This loss is particularly critical to their prospects because it takes the form of a balance-of-payments deterioration, inasmuch as economic growth is constrained by the availability of foreign exchange. And the loss is likely to be compounded insofar as the payments deterioration undermines their creditworthiness and thus curtails their ability to borrow. The IBRD has estimated that these effects are likely to prevent any significant growth in the per capita income of the 800 million people in the poorest of the developing countries for a decade.[18]

On the national level, an action which causes an important redistribution of income is often accompanied by offsetting action to protect those least able to bear the consequences. In the international case, an increase in grant aid of $10 billion would suffice to protect all the developing countries, while only $2 billion would be needed to compensate the poorest.[19]

It is doubtful whether there will be an increase in aid flows of this magnitude. It is true that Kuwait has pursued a rather generous aid policy for some years; that there are bonds of sympathy between members of OPEC and other developing countries which could be expected to induce other members of OPEC to follow Kuwait's example; and that information to date indicates that members of OPEC have been earmarking for aid sums which are substantial when measured by the standards used to evaluate Western aid programs.[20] But it is also true that much of this aid is being given as project aid through the medium of special funds and that disbursements are therefore likely to be subject to long lags; it seems unlikely that disbursements to all developing countries exceeded $3 billion in 1974. Far more fundamental, however, is the sheer arithmetic involved. Since OPEC is getting close to 15 percent of its additional revenue from the developing countries, and the increased oil revenue represents over 50 percent of current GNP for the countries that are sufficiently wealthy

18. International Bank for Reconstruction and Development, *Annual Report,* 1974, p. 5.

19. If increased aid took the form of loans, rather than grants, but the present value of the grant element amounted to $10 billion ($2 billion going to the poorest countries), the position of the developing countries (including the poorest among them) would be improved in comparison to the status quo ante, insofar as the shadow price of foreign exchange exceeds the official exchange rate. It may be reasonable, however, to ignore this second-order effect.

20. The flow of official economic assistance from OPEC to other developing countries from January to September 1974 amounted to $8.6 billion, excluding contributions to the IMF oil facility and more than $1 billion lent to the IBRD. See IMF *Survey,* 18 November 1974, pp. 357, 360–62.

to be potential donors of aid, the complete offsetting of the additional burden on the developing countries through increased aid by members of OPEC alone would require aid programs with a grant element of over 7.5 percent of GNP. This is more than ten times the objective proclaimed, let alone achieved, by the West. Moreover, unless the previous distribution of income was judged more appropriate than the new one, there is in logic no reason why the members of OPEC alone should be expected to bear the entire burden of the desirable increase in aid.

Several policy initiatives designed to increase the flow of aid have been taken by the international community during 1974. The IBRD has launched a drive to borrow some $2.5 billion per annum from the oil exporters and had raised some $1.15 billion by September 1974. The IMF has launched its oil facility and secured commitments by the oil exporters to lend some $3 billion; although this money is not described as aid, the facts are that a substantial proportion of the early loans made were to developing countries and that the interest rate is concessional. The United Nations launched its Special Fund, designed to relieve the hardest-hit of the developing countries by giving them grant aid, following the special session of the United Nations Conference on Trade and Development (UNCTAD) in April, with a target of $3 billion; to date the sums promised have been only $500 million from the European Community and $150 million from Venezuela.

After adding the aid that will be redirected away from members of OPEC, the project aid from the several development funds that were set up or expanded by the oil exporters, and the credit for oil imports given by certain members of OPEC to certain of the developing countries, these additional resources might in the aggregate come within striking distance of the balance-of-payments cost of the oil price increase. Since most of the additional funds are loans rather than grants, however, this would do little to offset the loss to the developing countries in real income but would merely prevent the effects of that loss from being compounded by an intensified balance-of-payments problem. Even that prospect may be unduly optimistic: Increasing indebtedness will still tend to erode creditworthiness and thereby curtail the scope for foreign borrowing. It seems likely, moreover, that some of the least-developed countries will fail to attract substantial additional aid from OPEC, so severe problems would arise from maldistribution of aid even if the aggregate level were adequate.

I have elsewhere[21] suggested that the prospect of securing agreement on the way in which the provision of aid should be shared between the members of the Development Assistance Committee (DAC) of the OECD on the one hand and the members of OPEC on the other would be promoted by explicit recognition that it is the "aid burden"—the present value of income sacrificed by the provision of aid—which should be divided between countries on the basis of some indicator of wealth, such as GNP. Such an approach would allow—and indeed encourage—donors of aid with strong payments positions but limited income levels, such as most of the members of OPEC, to provide a large volume of aid with a limited concessional element. Donors of aid with weak payments positions but high income levels, such as most of the traditional DAC donors at the present time, would concentrate on providing a smaller gross aid flow, but with a very high concessional element, to the least-developed countries. Specification of aid obligations along these lines might do something to insure that the problems caused to the developing countries by the loss of real income are not reinforced by a payments deterioration.

However, the worsening of the already grim prospects for the least fortunate will not be reversed without additional aid with a grant element comparable to the initial loss of real income. The sums required to safeguard the position of the least-developed countries are by no means vast. In the short run, one can but hope—if with no great optimism—that existing aid channels will be expanded piecemeal to deal with the increased need. In the longer term, one might hope that the shock to the global geoeconomic structure occasioned by the oil price increase might lead to a more far-reaching reform of the international economic system than would previously have been conceivable, in which an explicit commitment to secure a minimal degree of international redistribution of income along egalitarian lines would safeguard the position of the least fortunate. For it is the arbitrariness of the international income distribution, and the unwillingness of the international community to take coherent action to redistribute income, which the oil price increase has demonstrated to be the fundamental weakness of the existing international economy.

21. John Williamson, "More Flexibility in Meeting Aid Targets Could Raise the Value of Aid to Recipients," IMF *Survey,* 4 November 1974, pp. 351–53.

Trends in the International Oil Market

JOSEPH A. YAGER *and* ELEANOR B. STEINBERG

Brookings Institution

BARRING A NEW WAR or some other cataclysmic event, developments in the international oil market in the next two or three years will still be dominated by the changes set in motion by the Arab-Israeli war of October 1973 and the huge increase in oil prices in January 1974. An understanding of what has happened to the market during 1974 and early 1975 is therefore essential to any effort to analyze possible future trends. But before going into these recent events, it is useful to set them in perspective by reviewing briefly the principal developments affecting the market in the years before the war and to recall patterns and trends in energy production and consumption as they were before the war.[1]

Review of Past Developments

After having risen steadily during World War II, market prices of crude oil fell sharply in the late 1940s. Prices were relatively stable during most of the 1950s, but in the late 1950s and early 1960s increased production

1. The review of events up to the end of 1973 and of past energy trends that is presented here is based principally on Chapters 2 and 13 of *Energy and U.S. Foreign Policy* by Joseph A. Yager, Eleanor B. Steinberg, and others (Ballinger Publishing Company, 1974).

THE AUTHORS are grateful for information provided by Robert De Bauw, M. D. Gallard, Koji Hirota, Kenichi Matsui, and Arthur J. Ramsdell.

by independent producers, greater Soviet exports and the rapid exploitation of Libya's oil reserves ushered in a long period of excess supplies and
declining prices.

In 1960, the major oil-exporting nations formed the Organization of
Petroleum Exporting Countries (OPEC) with the purpose of preventing
any further reduction in the posted prices on which their taxes and royalties were based. They succeeded in this effort, but they were unable to
agree on the restrictions on production that were needed to stabilize
market prices.

The brief Arab-Israeli war in June 1967 marked the beginning of the
end of the buyers' market for oil. The embargo on shipments of oil to the
United States and Great Britain declared by several Arab countries was
both short-lived and ineffective, but the abrupt conversion of the Suez
Canal from a major channel of commerce to a military frontier contributed significantly to the rise of a strong sellers' market a few years
later. Failure of the two sides to agree on reopening the canal reinforced
the tendency of Western Europe to draw more and more of its oil from
North Africa, whose oil was attractive because of its low sulphur content
and its transportation advantage. Dependence on Libya was further increased by the civil war in Nigeria which interfered with oil production in
that country. In early 1970, oil supplies for Western Europe were suddenly
tightened by an unexpected rise in consumption and by Syria's disruption
of the pipeline (Tapline) that had carried part of Saudi Arabia's oil to
the Mediterranean for transshipment to Europe.

The new Libyan revolutionary government was quick to realize the
critical importance of Libyan oil to Western Europe. The Libyan authorities first tightened the market still further by ordering production cutbacks
in the name of conservation. They then demanded that the oil companies
increase both posted prices and tax rates and threatened to shut down
production altogether if their demands were not met. By September 1970,
all the companies operating in Libya had given in to the government's
demands.

The Libyan breakthrough caused other oil-exporting countries to
demand new agreements with the companies. In December 1970, OPEC
passed a resolution calling for a "uniform general increase in oil prices"
and threatened "concerted and simultaneous action." In February 1971,
representatives of the major oil companies and the exporting countries

along the Persian Gulf met in Tehran. They agreed to increase posted prices about 35 cents a barrel (to be raised an additional 5 percent annually until 1975) and the tax rate to 55 percent. In April, at a meeting in Tripoli, the companies gave Libya a tax rate of 55 percent also, but raised the posted price of Libyan oil 90 cents a barrel, to take account of its lower sulphur content and the shorter distance from Libya to Europe. During the first part of 1971, other oil-exporting countries achieved similar gains through negotiation or unilateral action.

The devaluation of the U.S. dollar in August 1971 precipitated another, more modest round of price increases. In January 1972 in Geneva, the companies granted the oil-exporting countries an additional 8.5 percent increase in posted dollar prices.

The inelasticity of demand for petroleum products permitted the companies to pass on to consumers much of the increase in the payments that they were required to make to the oil-exporting countries. By the summer of 1973, the price of 34° Arabian light,[2] f.o.b. Persian Gulf, was approximately $2.70 a barrel, more than 80 percent higher than the price (in constant dollars) that prevailed at the beginning of 1970.

This price increase—which in retrospect appears modest—was a source of concern in the major oil-importing countries in the period immediately preceding the October 1973 war. Power in the international oil market was clearly slipping away from the once dominant international oil companies and into the hands of the governments of the oil-exporting countries. Neither the companies nor the governments of the major oil-importing countries were sure what the future would bring or what their policies should be in a rapidly changing situation.

The uncertainty and uneasiness of the companies and the oil-importing governments were increased by the concurrent increase in pressure by the oil-exporting countries for control over the oil deposits and production facilities within their borders. In some countries, this pressure took the form of outright nationalization; in others, the companies were asked to enter into participation agreements under which the host government would acquire majority control of oil company assets in the course of a period of years. In either case, the companies found that they had no effective means of resistance.

2. Oil prices cited in this chapter refer to this "benchmark" or "marker" crude.

Prewar Energy Trends and Prospects

During the 1960s, total consumption of energy in the non-Communist nations grew at the rate of 5.4 percent a year. Before the October 1973 war, it appeared that this rate of growth would be exceeded somewhat in the 1970s. Oil consumption increased at an average annual rate of 7.6 percent between 1960 and 1970. Despite the rapid growth foreseen for nuclear energy, oil consumption was expected to rise at the still steep rate of 6.9 percent annually during the 1970s; it was expected nearly to double between 1970 and 1980, and the share of total energy consumption that it represents was projected to rise from about half to about 60 percent.

For most countries, the anticipated rise in oil consumption also meant a rapid rise in oil import requirements. Increases in domestic oil output in the net importing countries were expected to meet only a small part of the total increase in demand. Thus, before the October 1973 war, it appeared quite possible that in 1980 oil imports would reach 48 million barrels a day—an increase of almost two-thirds over total imports in 1972. There was little doubt that the oil-exporting countries possessed the physical resources needed to meet these requirements. In fact, it appeared that even with this continued growth in the market, at least some of those countries would have to exercise restraint in expanding output, if excess production and the return of a buyers' market were to be avoided.

The October War and Its Aftermath

On the eve of the war, government officials and informed members of the public in the major oil-importing countries were concerned over the future of the international oil industry, but—as events would soon show— they were most worried about secondary problems. Considerable attention was devoted to the crumbling power of the major oil companies and what that might portend for the future functioning of the industry and for the roles of governments in setting the terms under which oil would be produced and marketed. Some observers in oil-importing countries feared that the oil-exporting countries would use their new market power to force a continued rise in prices, but no one foresaw the huge upward leap in prices that was soon to take place.

Some thought was given in the major oil-importing countries to the possibility that the Arab oil exporters would declare some kind of an embargo to demonstrate their impatience with the failure of the big powers to force the Israelis to give up the land that they overran in 1967. The possibility of an effective Arab embargo was not rated very high, however, in the light of the dismal failure of the Arabs to make good use of oil as a weapon in past crises. Few people outside the Arab world thought that the Arabs would initiate another round of fighting in the face of what was assumed to be the overwhelming military superiority of Israel.

At first, the surprise Arab attack of 6 October 1973 was, with good reason, seen in world capitals primarily as a serious threat to the peace. Only gradually did oil move to the center of the stage.

On 16 October, the Persian Gulf oil-exporting countries—including non-Arab Iran—unilaterally announced an increase in the posted price of oil that had the effect of raising the market price, f.o.b. Persian Gulf, to $3.65 a barrel, nearly a dollar above the previously prevailing level. Other exporters followed suit, and in the uneasy atmosphere created by the war, the market readily absorbed the increase.

The troubles of the oil-importing countries had, however, only begun. On 21 October the Arab oil-exporting countries proclaimed an embargo on shipments to the United States, the Netherlands, and several other countries regarded as being too sympathetic to Israel. More important, all the Arab oil-exporting countries except Iraq simultaneously cut oil production by 5 or 10 percent and announced that deeper cuts would be made unless their objectives with respect to Israel were realized. On 4 November participating Arab countries announced that their production would be cut to 75 percent of the September level.[3]

Part of the reduction was of course to be absorbed by the United States and other "unfriendly" countries that were totally denied Arab oil. The remainder was to come out of exports to countries that the Arabs regarded as neutral. The United Kingdom, France, and other "friendly" countries were assured of normal supplies. What actually happened was that the impact of the Arab supply restrictions was felt more evenly by all oil-importing countries than the Arabs had intended, because the oil com-

3. Total Arab production actually fell to about 80 percent of the September level in both November and December, or by about 3.4 million barrels a day. See Federal Energy Administration, Petroleum Industry Monitoring System, *U.S.-OPEC Petroleum Report,* Year 1973, 1 July 1974.

panies shifted non-Arab oil to customers who had been denied all or part of what they would normally have imported from Arab countries.

For this reason, and because in late December the Arabs raised production to 85 percent of the September level, the oil shortages arising out of the Arab resort to their oil weapon were much less severe than had at first been feared. Much more serious and lasting in its effects was the great increase in oil prices that the Arab supply restrictions made possible.

In late December 1973, the Persian Gulf producers again took the lead and more than doubled the posted price of crude oil, effective 1 January 1974. The average market price appeared as a consequence to have been raised to about $8.20 a barrel, f.o.b. Persian Gulf, three times what it had been only a year earlier.[4] Comparable price increases were soon made by the other exporting countries.

Developments during 1974 and Early 1975

The best way to obtain an overview of developments affecting the international oil market since January 1974 is to review separately the policies pursued by the oil-exporting and oil-importing countries.

Policies of the Oil-Exporting Countries

As the year 1974 opened, uncertainty over the way in which the Arabs would use their oil weapon still dominated the international oil market. This political factor, however, rapidly gave way to considerations of a more purely economic nature. Pursuant to a decision taken in late December 1973, the Arabs relaxed their supply restrictions during January and February. In March, they removed them altogether along with the embargo on shipments to the United States. The issue of overriding importance to all oil-exporting countries, Arab and non-Arab alike, then

---

4. This estimate rested on the assumption—which was justified at the time—that equity crude would account for 75 percent of all oil shipped and that its tax-paid cost would be $7.00 plus 35 cents for production costs and company profits. It was further assumed that participation crude (the other 25 percent) would be sold for 93 percent of the posted price. Later in 1974, however, retroactive application of an increase in the share of participation crude to 60 percent of the total caused the effective 1 January price to be substantially higher than had been at first assumed (see Table 7-1).

became what to do about prices. Should the high prices established on 1 January be maintained or pushed even higher, and, if so, how? If not, when and by how much should prices be reduced?

The issue was of course rarely posed in such simple terms. More often than not, the question to be decided in OPEC's price deliberations was whether the posted price (not the market price) of equity crude, the oil owned by the foreign oil companies, should be changed. At other times, the subject under debate was whether the taxes and royalties paid by the companies to the oil-exporting governments on equity crude should constitute a higher percentage of the posted price. The problem was further complicated by the fact that the revenues of the oil-exporting governments depended not only on the taxes and royalties that they received on equity crude, but also on what they were able to get for participation crude, the oil that they owned under the terms of the participation agreements.[5] This depended in turn principally on the buy-back provisions of those agreements, under the terms of which the companies were permitted (or required) to purchase the participation crude that the governments could not market by other means. In early 1975, the proportion of participation crude marketed directly by the governments concerned was still quite small.

During 1974, the buy-back price of participation crude was 93–94 percent of the posted price, substantially above the tax-paid cost of equity crude. This meant that, even if nothing else changed, an increase in government participation rights would yield an increase in oil revenues. This is what in fact happened during 1974 as more governments demanded and obtained an increase in their participation shares from 25 percent to 60 percent. Since the increase in participation shares was applied retroactively to 1 January, the effect of the rise in the posted price on the proceeds to governments from the average barrel of crude—and therefore on the market price—was much greater than had at first been supposed.

The relation between participation shares on the one hand and government revenues and prices on the other must be kept in mind in evaluating developments in the international oil market since 1 January 1974. If

5. These agreements had become the most common arrangement between governments and companies in the Arab Middle East. Other arrangements existed elsewhere, but in most cases a distinction was drawn between the part of total oil production that was normally to be marketed by the companies and the part that was wholly at the disposal of the governments.

Table 7-1. *Estimated Market Prices of Saudi Arabian Light Crude Oil in 1974*

U.S. dollars a barrel, f.o.b. Persian Gulf

Date	Price
January 1	9.66
July 1	9.77
October 1	10.10
November 1	10.47
December 31	10.47

Source: Estimated market prices are based on figures for government proceeds in the *Petroleum Economist* Vol. 42 (February 1975), p. 71. Sixty percent of the oil marketed is assumed to have been participation crude and forty percent equity. Ten cents has been added for production costs and twenty-five cents for company profits. The latter figure is low in comparison with past norms, but it is believed to be appropriate for much of 1974.

account is taken only of what effective prices have turned out to be after the increase in participation shares from 25 percent to 60 percent, the price of crude rose only from $9.66 to $10.47 a barrel during 1974, a relatively modest 8.4 percent. (See Table 7-1.) If, however, the price that appeared to have been established on 1 January 1974 is taken as the base, the twelve-month increase was from $8.20 to $10.47, an impressive 28 percent.

From the latter perspective, the debate over price policy within OPEC during 1974 and early 1975 loses some of its importance. The price of crude was being influenced more by decisions on participation rights than by OPEC's fine tuning of tax and royalty rates. Nevertheless, OPEC's deliberations on price policy are of some interest, since they may provide some clues to possible future developments.

On the public record, the debate over price policy within OPEC appears to have been between the Saudi Arabians, who advocated somewhat lower prices, and all the other members, who favored either holding the line or pushing prices still higher. The Saudis argued publicly that existing oil prices threatened the stability of the international economy, but as the possessors of the world's largest oil reserves, they were presumably also concerned that high oil prices would stimulate the development of other sources of energy. The OPEC majority justified its position by pointing to the profits of the oil companies and to the rising cost of the commodities that they imported from the industrialized countries.

In June 1974, OPEC ministers meeting in Quito raised royalties by 2 percent or about 11 cents a barrel, effective July 1. The Saudi representative, Sheikh Ahmed Zaki Yamani, is reported to have opposed even this small increase and pressed unsuccessfully for lower prices.[6] At the September 1974 OPEC meeting in Vienna, Saudi Arabia again urged a cut in oil prices and publicly refused to go along with the decision of the other members of OPEC to increase proceeds to governments 3.5 percent, or about 33 cents a barrel, effective 1 October.[7] On this occasion, however, Saudi Arabia was reported to be uncomfortable in its lone role as an advocate of lower prices.[8] During the Vienna meeting, it also became known that three months earlier Saudi Arabia had itself initiated a price increase by setting the buy-back price for participation crude at 94.864 percent of the posted price. The companies reportedly had been under the impression that the rate would be 93 percent.[9]

After the Vienna meeting, consumers would have been justified in believing that no further increases in the price of oil would be imposed in 1974. After all, at that meeting the OPEC ministers agreed that crude prices would be frozen for the remainder of the year; higher payments by the companies, it was asserted, could come out of their "excessive profits."[10] On 9 and 10 November, however, the Persian Gulf oil producers (with the exception of Iran) met in Abu Dhabi to consider a new pricing formula proposed by Saudi Arabia. At the close of the meeting, Saudi Arabia, the United Arab Emirates, and Qatar agreed to lower posted prices and to increase tax and royalty rates. The net effect of these changes was to raise the proceeds to governments on the average barrel of oil by 37–50 cents[11] and to narrow the difference between the cost to the companies of participation crude and equity crude.

One month later, the OPEC ministers meeting again in Vienna endorsed the Saudi pricing formula and declared that for the first nine months of 1975 the weighted average proceeds from operating companies for Arabian light marker crude would be $10.12 a barrel.[12] If 35 cents is added for production costs and company profits, the resulting average

6. *Wall Street Journal,* 18 June 1974, p. 2; *New York Times,* 18 June 1974.

7. *Wall Street Journal,* 16 September 1974, p. 3; *Petroleum Economist,* Vol. 41 (October 1974), p. 362.

8. *New York Times,* 17 September 1974.

9. *Wall Street Journal,* 17 September 1974, p. 5.

10. *Oil and Gas Journal,* 21 September 1974, p. 108.

11. *Economist,* 16 November 1974, p. 112, and the *Oil and Gas Journal* newsletter, 18 November 1974, use the 50-cent figure. The *Petroleum Economist,* Vol. 42 (February 1975), p. 71, appears to have calculated the increase at 37 cents.

12. *Oil and Gas Journal,* 23 December 1974, p. 15.

market price is $10.47 a barrel. The decision to hold oil prices at this level through September 1975 was reaffirmed at the meeting of OPEC ministers in Libreville, Gabon, in June. At the same time, however, OPEC declared its intention "to readjust crude oil prices as from October 1, 1975" and to link oil prices to the International Monetary Fund's special drawing rights (SDRs).[13]

Coming on top of a slowdown in economic activity throughout the industrialized world, the high price policy of OPEC contributed to a stagnation in the demand for oil during 1974. Total consumption of oil in the non-Communist nations (not including the oil-exporting countries) in that year was about 42.2 million barrels a day, 3.9 percent below 1973 consumption (see Table 7-9). The oil-exporting countries therefore had to face the very real possibility that production would outrun demand and force prices down.

This problem did not of course arise during the first quarter of the year, when the Arab supply cutback was still holding nearly 3 million barrels of oil a day from the market.[14] Even after the supply restrictions were lifted, the Arab oil-exporting countries as a group did not immediately restore production to the level that had been reached before the war. In June 1974, total Arab production was still only 93 percent as large as the output of September 1973.[15] In the cases of Libya and Kuwait, this restraint was the result of publicly proclaimed conservation policies.[16]

Not surprisingly, most of the major non-Arab oil-exporting countries —Iran, Indonesia, and Nigeria—increased their production during the first half of 1974 over that of the same period in 1973. A notable exception was Venezuela, whose production was reduced by 7.4 percent, partly in order to conserve natural gas that is flared (burned off) at producing oil wells.[17] Nevertheless, production of oil by the non-Arab exporters went up only about 4 percent during the first six months of 1974 (see Table 7-2). Despite the restraint implicity in this figure, oil production in mid 1974 was outrunning current consumption by an amount variously estimated at 2 million to 4 million barrels a day. Storage capacity was soon full, and further cuts in output became necessary. Kuwait and Venezuela

13. *New York Times,* 12 June 1975.
14. *Petroleum Economist,* Vol. 41 (January 1974), p. 6.
15. *Oil and Gas Journal,* 26 August 1974, pp. 46, 161.
16. *New York Times,* 21 March 1974.
17. *Wall Street Journal,* 8 April 1974, p. 4.

Table 7-2. *Estimated Crude Oil Production in Non-Communist Nations,
1973 and 1974*

Millions of barrels a day

Region and country	First half 1973	Second half 1973	First half 1974	Second half 1974
Western Hemisphere				
United States	10.3	10.4	10.0	9.8
Canada	2.0	2.0	2.0	1.9
Venezuela	3.5	3.6	3.2	3.0
Other	1.7	1.8	1.8	1.9
Middle East				
Saudi Arabia	7.4	7.8	8.3	8.8
Iran	5.8	6.0	6.1	6.0
Kuwait	3.0	3.1	2.9	2.2
Iraq	1.9	2.2	1.9	1.9
Abu Dhabi	1.3	1.2	1.4	1.3
Other	1.6	1.6	1.5	1.6
Africa				
Nigeria	1.9	2.1	2.2	2.3
Libya	2.2	2.0	1.8	1.3
Algeria	1.0	1.0	1.0	0.9
Other	0.4	0.5	0.5	0.6
Western Europe	0.3	0.3	0.3	0.3
Far East				
Indonesia	1.2	1.4	1.4	1.4
Other	0.8	0.9	0.9	0.8
Total	46.6	47.9	47.4	46.0

Source: *Petroleum Economist*, Vol. 41 (September 1974), p. 328, and Vol. 42 (January 1975), p. 10.
Detail may not add to totals because of rounding.

announced that oil production would be reduced to help adjust supply to demand. Other oil-exporting countries may have quietly taken similar steps.[18] A large part of the cutback was in effect imposed by the oil companies, which were unable to market all the oil being produced at prices that would cover the cost of crude.[19]

Total oil production in the second half of 1974 was 4 percent below that of the second half of 1973, despite the fact that production in the latter period was affected by the Arab supply cutbacks. Certain shifts in the market shares of various nations are also potentially significant (see

18. *Oil and Gas Journal,* 2 September 1974, pp. 33–34.
19. *Wall Street Journal,* 28 August 1974, p. 3.

Table 7-3. *Market Shares of Principal Non-Communist Oil-Producing Countries, 1973 and 1974*

Percent

Country	First half 1973	Second half 1973	First half 1974	Second half 1974
United States	22.1	21.7	21.1	21.3
Canada	4.3	4.2	4.2	4.1
Venezuela	7.5	7.5	6.8	6.5
Saudi Arabia	15.9	16.3	17.5	19.1
Iran	12.4	12.5	12.9	13.0
Kuwait	6.4	6.5	6.1	4.8
Iraq	4.1	4.6	4.0	4.1
Abu Dhabi	2.8	2.5	3.0	2.8
Nigeria	4.1	4.4	4.6	5.0
Libya	4.7	4.2	3.8	2.8
Algeria	2.1	2.1	2.1	2.0
Indonesia	2.6	2.9	3.0	3.0
Other	10.3	10.6	10.6	11.3

Source: Table 7-2. Percentages may not add to 100 because of rounding.

Table 7-3). The largest gains were made by Saudi Arabia, Nigeria, and Indonesia. Iran also recorded a modest increase in its share of the market. The big losers were Libya, Kuwait, and Venezuela. Iraq's failure to expand its share of the market is also notable, in the light of its proclaimed desire to do so.[20] Since these shifts in market shares were the unco-ordinated result of actions taken by individual governments and companies for a variety of reasons, it will be no surprise if the new division of the market proves not to be stable.[21]

The decline in the production of oil continued in the first quarter of 1975. Total output in non-Communist areas was 11.3 percent below that of the first quarter of 1974, when production was affected by the

20. *New York Times,* 23 December 1974.

21. One sign of the strains that may exist on the producers' side of the market was the controversy that broke out in February 1975 between Abu Dhabi and its operating companies. The Petroleum Minister of Abu Dhabi complained publicly both about the decision of operating companies to cut production in the sheikhdom from 1.2 million barrels a day in December to only 0.7 million barrels a day in February and about the failure of OPEC to co-ordinate production by its members in order to avoid surpluses. He ordered production restored to the December level (*Washington Post,* 21 February 1975). But in return, Abu Dhabi agreed to cut the price of its crude oil by 55 cents a barrel, virtually eliminating the premium previously charged for low sulphur content (*Wall Street Journal,* 3 March 1975, p. 7). This reduction was subsequently approved by OPEC as a special case.

Arab supply restrictions. Of particular interest were the reductions in output in Saudi Arabia and Iran, countries in which there had been no serious decline in output during 1974.[22]

Policies of the Major Oil-Importing Countries

The initial reactions of the major oil-importing countries to the Arab embargo and supply restrictions were unco-ordinated and of limited effectiveness. The European members of the OECD were unable to agree to activate their existing emergency allocation plan, presumably because nations on the Arabs' "friendly" list feared losing their preferred status, and nations in the neutral category wished to avoid being labeled as hostile. Broader co-operation among all members of the OECD was widely discussed, but without practical result so far as meeting the immediate crisis was concerned.

The Washington Energy Conference, convened by the United States government in February 1974 and attended by representatives of Japan, Canada, and ten European nations, set in motion a process of consultation that may lead to a co-ordinated energy policy for most of the major oil-importing nations. The conferees (with the exception of the French) agreed on the need for a comprehensive program of action that would concert national policies in such areas as conservation of energy, allocation of oil supplies in times of emergency, development of additional energy sources, and energy research and development.[23] Subsequent consultations among the major oil-importing countries led in November 1974 to the establishment of a sixteen-member International Energy Agency (IEA) closely associated with the OECD.[24]

Acting on a U.S. initiative, the IEA developed a complicated emer-

22. *Oil and Gas Journal* newsletter, 19 May 1975.

23. U.S. Department of State, Washington Energy Conference communiqué, Doc. 17, Rev. 2 (13 February 1974).

24. The original members were Belgium, the Netherlands, Luxembourg, Germany, Italy, Denmark, Ireland, the United Kingdom, Spain, Sweden, Switzerland, Austria, Turkey, Canada, the United States, and Japan. New Zealand joined later, and Norway became an associate member. France has not joined, but maintains a close co-operative relationship with the new organization. See the *New York Times,* 16 November 1974 and 6 February 1975; see also a statement by Assistant Secretary of State Thomas O. Enders before the Senate Committee on Interior and Insular Affairs, 13 February 1975, in U.S. Department of State news release, February 1975, cited hereafter as Enders statement.

gency program which—when ratified—would commit member governments to the building of common levels of emergency oil stocks, to the development of "pre-positioned demand-restraint programs" that would enable them to cut oil consumption by agreed-upon common rates during emergencies, and to the allocation of available oil (both domestic production and continuing imports) during emergencies.

By mid March 1975, the oil-importing nations, working either within the IEA or within the broader framework of the OECD, had agreed in principle on three other major actions.[25]

• Reduction of oil consumption through conservation measures.

• Creation of a $25 billion safety net to help any of their number encountering serious balance-of-payments difficulties because of the high price of oil. Borrowers would be required to "show that they are making a strong effort in conjunction with other IEA members to conserve energy and develop new energy sources."[26]

• Establishment of a common floor price for oil within each of the IEA nations.

With respect to each of these there were details—some of them important —that remained to be worked out before the formal approval of governments would be sought. .

The impact of these measures on the international oil market, if they should be put into effect—and it is by no means certain that they will—is difficult to estimate. How much oil is to be saved over what period of time through conservation is not yet clear.[27] The showing of energy conservation and development required of governments resorting to the financial safety net could have a significant effect on the oil market, but whether this requirement would be enforced rigorously may be questioned. A floor price for oil was initially advanced as a means of encouraging continuing investment in domestic sources of energy by providing protection against a substantial fall in the international price of oil.[28] How strong this en-

25. *New York Times,* 21 March 1975; Enders statement.

26. Enders statement.

27. The Governing Board of the IEA has set a goal of reducing oil imports of its members 2 million barrels a day by the end of 1975 ("International Aspects of the President's Energy Program," U.S. Department of State news release, 11 March 1975, p. 2). Goals have not yet been set for future years, and the 1975 goal is subject to various interpretations.

28. A common floor price for all IEA members is now seen primarily as a means of keeping nations embarking on high-cost energy projects from being placed at a competitive disadvantage.

couragement would be depends, however, on the level of the floor price, which is still an issue within IEA. In any event, because of the long lead times in energy-development projects, adoption of a floor price could not be expected to have much effect on energy supplies in the near future.

Agreement in principle on the U.S. energy proposals was reached as soon as it was only because the United States insisted on substantial progress toward a common energy policy among the oil-importing countries before it would agree to attend the preparatory meeting held for the purpose of planning a conference between oil-importing and oil-exporting nations. This conference, proposed by France in October 1974, was finally convened in Paris in April 1975. The meeting was attended by representatives of the industrialized oil-importing countries (the United States, the European Community, and Japan), the oil-exporting countries (Algeria, Saudi Arabia, Iran, and Venezuela), and the oil-importing developing countries (Brazil, India, and Zaire). Nine days later the delegates disbanded without agreeing on an agenda for a plenary conference that was to have been scheduled to meet in July or August. The developing countries—both oil-importing and oil-exporting—insisted that the plenary conference consider a wide range of international economic issues. The industrialized countries held out for an agenda on which primary emphasis would be given to energy-related problems.[29] Efforts to resolve these differences continued after the failure of the preparatory meeting,[30] however, and a plenary conference of oil-importing and oil-exporting nations may yet be held.

Efforts by the European Community (EC) to develop a common energy policy have been overshadowed by developments in the broader forum provided by the OECD and the new IEA. The statement on energy policy approved by the EC foreign ministers in September 1974 did not go beyond the setting of general guidelines advocating the reduction of fuel imports from outside the Community, more efficient use of Community energy resources, development of a plan for meeting energy shortages, and common energy research efforts, especially in the field of nuclear energy.[31] Relations between the EC and the Arab countries improved somewhat following the lifting of the embargo, but efforts by the EC to engage in a constructive dialogue with the Arab League on possibilities

29. *New York Times,* 17 April 1975 and 15 July 1975.
30. *Wall Street Journal,* 28 May 1975, p. 3.
31. *New York Times,* 18 September 1974.

for economic, technical, and cultural co-operation have apparently achieved very little.[32]

The unilateral actions taken by individual oil-importing governments during 1974 and early 1975 to deal with their energy problems have been principally of two kinds: efforts to conserve oil and efforts to insure the future availability of imported oil. How domestic production of energy materials might best be increased has been the subject of intensive discussion and some action, but because of the complexity of the problem and the long lead times involved, few significant results have been achieved. The blueprint for Project Independence reached President Ford's desk only in November 1974, a full year after the project had been launched by his predecessor.

Oil conservation efforts by governments in response to the embargo were in most cases quite modest—tighter allocation systems, lower speed limits, carless Sundays, exhortations on the need for conservation, and the like. If the Arab supply reductions had been deeper or more lasting (or if the winter of 1973–74 had not been unusually mild) more drastic actions would clearly have been required. Many of the oil-saving measures that were adopted during the embargo were allowed to lapse, in some instances before the embargo was lifted.[33] The fact that oil consumption did not rise during 1974 must be attributed much more to the sluggish state of the economies of the industrialized nations and to consumer reactions to higher oil prices than to government conservation programs.

This picture may of course change in the next few years. Several oil-importing countries are developing longer-range oil-conservation programs. For example, France set a ceiling on the amount to be spent on oil imports in 1975[34] and raised energy prices.[35] Britain increased taxes on gasoline and imposed a speed limit of 50 miles an hour on most roads.[36] West Germany ordered a 10 percent increase in consumption of coal in place of oil by electric power stations.[37] Italy, despite its domestic political problems, announced a program intended to reduce oil consumption by 7 percent.[38]

32. *Economist,* 24 May 1975, p. 33.
33. *New York Times,* 14 October 1974.
34. *Wall Street Journal,* 26 September 1974, p. 21. See also *Petroleum Economist,* Vol. 42 (April 1975), p. 126.
35. *Wall Street Journal,* 2 January 1975, p. 4.
36. *New York Times,* 10 December 1974.
37. *Petroleum Economist,* Vol. 42 (January 1975), p. 29.
38. *New York Times,* 4 February 1975.

In October 1974, President Ford proclaimed the goal of cutting oil consumption one million barrels a day by the end of 1975.[39] The President subsequently announced that phased increases in the fees on imported oil would be a major means of achieving this goal, and on 1 February 1975, he raised the fee per barrel of crude oil by $1.[40] On 1 June, he raised the fee on imported crude oil an additional $1 a barrel and increased the fee on imported refined petroleum products 60 cents a barrel. At the same time, he announced his intention to phase out control of the price of domestic crude oil. Both measures, however, are strongly opposed by some members of Congress.[41] In July 1975, it was still not clear what measures Congress and the President would settle upon to achieve the reduction in oil consumption that both sides agreed was necessary.

Efforts by various oil-importing nations to insure the future availability of imported oil ranged from broad schemes for economic and other co-operation with oil-exporting countries to explicit barter deals. In this respect, the most active oil-importing countries were probably France and Japan, although lesser efforts were made by the United Kingdom, Italy, and West Germany. The most sought-after partners from the ranks of the oil-exporting countries were Saudi Arabia, Iran, and Iraq; bilateral arrangements were, however, also made or discussed with Kuwait, Libya, and Indonesia. Only the United States took a firm stand against explicit barter deals, on the ground that they tend both to undermine the multi-lateral trading system and to drive up the price of oil.

The full story of the scramble by oil-importing countries for special bi-lateral arrangements with oil-exporting countries is quite complicated, and hard facts are in many cases unavailable. There is no reason, however, to believe that such arrangements have provided any of the oil-importing countries concerned with cheaper supplies of imported oil, or with iron-clad guarantees that supplies will not be restricted in a future crisis. The most that may have been achieved is an increase in the stake of both parties in maintaining good economic relations, thereby decreasing some-what the risk of future interruptions of oil supplies.

The United States in particular has been engaged in an effort to create an increasingly strong and complex network of mutual interest with the major oil-exporting nations. In addition, the United States has tried both quiet persuasion and public exhortation in an effort to bring oil prices

39. *Wall Street Journal,* 9 October 1974, p. 2.
40. *New York Times,* 24 January 1975.
41. *Wall Street Journal,* 28 May 1975, p. 3.

down. Saudi Arabia, the principal target of the U.S. diplomatic effort, made gratifying declarations of its belief that prices were too high, but took no actions that might have caused prices to come down. The use of public exhortation by the United States reached its peak in late September 1974, when President Ford and Secretary of State Kissinger delivered gloomy, vaguely threatening, speeches on the dire consequences of continued high oil prices.[42]

The Situation in Mid 1975

Halfway through the second year of the regime of high oil prices, the international oil market was dominated by uncertainty concerning the future. For more than a year, the market had been essentially in a holding pattern that provided few clues to what the next year or so would bring.

For the first time in many years, the world market for oil did not grow, but this phenomenon could be explained largely by the sluggish behavior of the economies of the industrialized nations. Consumer reactions to high oil prices also contributed to the stagnant state of the market. Government conservation programs had some effect during the first quarter of 1974, but their importance as a determinant of consumption probably diminished in later months. Too little time had passed for high oil prices to stimulate any significant increase in the availability and use of energy from other sources.

The chief unanswered question on the demand side of the market in the next few years is whether an upturn in the world economy will cause the international market for oil to expand significantly, even at current high prices, or whether the expansionary impetus to demand provided by an increase in economic activity will be blunted by more effective governmental conservation programs, continued consumer resistance to high prices, and an increase in the availability of substitutes for oil.

The major unanswered question on the supply side of the market is whether the informal adjustment mechanisms that worked after a fashion during 1974 and early 1975 will continue to work in the future. In particular, if the size of the international oil market contracts, the oil-exporting countries may have to decide whether to try to agree upon a system of rationing production or to allow the companies to decide where the reductions should be made.

42. *Wall Street Journal,* 26 September 1974, p. 2.

The Demand for Energy in the Oil-Importing Countries, 1975–77*

Before considering the various market strategies open to OPEC and its members over the next two or three years, it is useful to ask what may happen to the international oil market if the price of oil (in real terms) is held at the 1974 level through 1977. The answer will depend principally on the levels of economic activity in the industrialized non-Communist countries and on the continuing impact on consumer behavior of the huge oil price increases of late 1973 and early 1974.

The analysis which follows assumes that there will be no resumption of large-scale hostilities in the Middle East.

Base-case forecasts of economic activity in the major non-Communist regions used for projecting energy demand are shown in Table 7-4, below. Given the uncertainties and rapid changes that tend to characterize short-range economic forecasting, a range delineated by real economic growth rates of two percentage points below (Case I) and two percentage points above (Case II) the base forecast is shown for 1976 and 1977. These alternative assumptions concerning GNP will be used later in this section to provide a plausible range of estimates of total energy requirements and, ultimately, of total world oil import requirements for 1975 through 1977.

The overall picture suggested by the base-case projections is that of a weak performance in most of the industrialized world in 1975, followed by a recovery of uncertain proportions in 1976 and further improvement in 1977. The U.S. base forecast incorporates the expected stimulus from the $23 billion tax cut enacted in March 1975. It is difficult to know with a high degree of certainty, however, how fast the tax cut will result in positive growth and to project the dimensions of the recovery. The base forecast for Japan assumes the adoption of some policy measures to expand the supply of money. The forecast for Western Europe suggests that that region is experiencing a less severe recession than is the United States or Japan.

The traditional method used to forecast the demand for energy is based in part on the historical statistical correlation between energy consumption and economic activity—that is, a change of 1 percent in the rate of real economic growth is associated with a change of roughly 1 percent in

* Prepared with the assistance of Aliou B. Diao and Aeran Lee.

Table 7-4. *Real GNP Growth Rates for Major Industrialized Areas, 1973–77*
Percentage change from preceding year

Country or region	Actual		1975 Base case	Projected					
				1976			1977		
	1973	1974		Case I	Base case	Case II	Case I	Base case	Case II
United States	5.9	−2.1	−3.5	4.5	6.5	8.5	5.5	7.5	9.5
Western Europe	5.4	2.3	2.2	2.2	4.2	6.2	3.4	5.4	7.4
Japan	10.2	−1.8	4.1	3.3	5.3	7.3	5.5	7.5	9.5

Sources of base forecasts:
United States: George L. Perry, Brookings Institution.
Western Europe: OECD, *Main Economic Indicators,* July 1975, p. 156, for 1973 and 1974. Estimates for 1975, 1976, and 1977 are derived from the discussion of Western Europe in Chapter 3, by Giorgio Basevi.
Japan: OECD, *Main Economic Indicators,* July 1975, p. 156, for 1973 and 1974. For 1975 and 1976, Japan Economics Research Center, cited in *Japan Economic Journal,* 4 February 1975, p. 11. Estimate for 1977 by Brookings staff.

energy demand. However, using this correlation in forecasting short-run energy consumption involves serious problems. Historical data show that this ratio—known as the long-term GNP elasticity of demand for energy —varies widely among different countries and within any one country at different times.[43] Moreover, this ratio gives at best a rough indication of the way in which energy consumption might behave if the growth of GNP follows long-term trends. The effects of short-term deviations from trend —such as are occurring during the current recession—must be allowed for separately.

A regression analysis of annual percentage changes in real GNP and energy consumption in the United States in the period 1953–73 suggests

43. For example, for the period 1965–71, the average annual energy consumption/GNP elasticity for the United Kingdom was 0.64, for the United States it was 1.45, and for Japan, 1.05. The differences in the elasticity for a given country, moreover, changed significantly over a period of time, reflecting changes in the economic structure of the country. Thus, the ratio of energy consumption to GNP in the United States was 0.81 for the period 1960–65; the average ratio for the period 1965–71 was 1.45. For several other industrialized countries, the ratio declined significantly over the same periods of time. For purposes of this analysis, it is assumed that the economies of the industrialized countries will not undergo fundamental structural changes during the next two or three years. See Joel Darmstadter and S. H. Schurr, "World Energy Resources and Demand," in *Philosophical Transactions of the Royal Society of London,* 1974, pp. 276, 413–30, Table 9, p. 16. See also John G. Myers and others, *Energy Consumption in Manufacturing: A Conference Board Report to the Energy Policy Project of the Ford Foundation* (Ballinger Publishing Company, 1974), pp. 1–57.

that over the long run a change of 1 percent in GNP is associated with a change of a similar percentage in energy consumption. However, for every 1 percent of deviation of GNP from trend, the associated change in energy consumption is smaller than 1 percent.[44] (This conclusion is generally consistent with those of a number of other studies.[45])

A long-term GNP elasticity of demand for energy of 1.0 also seems appropriate for Western Europe and Japan.[46] The coefficient relating short-term deviations of GNP from trend with energy consumption should probably be somewhat greater for Western Europe than for the United States and considerably greater for Japan. This is assumed to be so because a much larger percentage of U.S. energy consumption goes for purposes (private transportation, air conditioning, home heating, and the like) that are not so tightly linked to the level of economic activity as is the case in the other major industrialized areas. Somewhat arbitrarily, this coefficient has been set at 0.7 for Western Europe and 0.9 for Japan.

Besides the off-trend effects of recession on energy consumption there is the question of the impact of the recent enormous price increases on the consumption of energy. (Other sources of energy for the most part are also rising rapidly in price in response to the 1973–74 oil price increase.) Before the large 1973–74 increases in the price of crude oil, very little research had been done on the price elasticity of demand for oil or energy. The principal example of the research in this area was a study by the U.S. National Petroleum Council, which was concerned with the long-term

44. If g is the potential (or trend) rate of GNP growth and a the actual rate, the percentage change in energy consumption, e, in a given year over that of the previous year attributable to changes in the level of economic activity can be derived from the following formula:

$$e = 1.0 \, (g) + 0.5 \, (a - g).$$

The authors are indebted to Charles L. Schultze for this formula and the underlying regression analysis.

45. See, for example, Joel Darmstadter, "Energy Consumption: Trends and Patterns," in Sam H. Schurr (ed.), *Energy, Economic Growth, and the Environment* (Johns Hopkins University Press for Resources for the Future, 1972), p. 196; see also National Petroleum Council, *U.S. Energy Outlook: Energy Demand* (1973), p. 42.

46. OECD, *Energy Prospects to 1985: An Assessment of Long Term Energy Developments and Related Policies,* A Report by the Secretary-General, 2 vols. (Paris: Organisation for Economic Co-operation and Development, 1974), Vol. 1, Tables 2-1, 2-5, and 2-7, pp. 43–53. (The energy consumption/GNP elasticities were derived from gross domestic product projections and the base-case energy demand shown in these two tables.)

(roughly fifteen years) price elasticity of demand for oil, and which analyzed much smaller price increases for the most part than those which occurred in 1973–74.[47] A number of studies have been undertaken recently on the short-term price elasticity of demand for oil in the United States.[48] Insufficient time has passed since the 1973–74 price increases, however, for an evaluation of the results of these recent studies to be possible.

Because of the unprecedented magnitudes of the price increases, which occurred after a relatively long period in the 1960s during which energy prices in real terms were declining, history offers little guidance in the matter. For purposes of the present analysis, what is assumed to be a very low price elasticity——0.12 over the four-year period 1974–77—was adopted for most of the forecasts presented below.[49] It is arbitrarily assumed that the price effects are spread out evenly during the four years— that is, that the incremental price elasticity each year would be −0.03.[50] What is thought to be a low price elasticity was selected in order to present a pessimistic case from the standpoint of the oil-importing nations. If the elasticity proves to be too low and consumption and imports are actually less than forecast, then the problems faced by the importers will clearly be easier to solve.[51]

47. National Petroleum Council, *U.S. Energy Outlook: Energy Demand*, Appendix E, pp. 79–88.

48. See, for example, E. A. Hudson and D. W. Jorgensen, "U.S. Energy Policy and Economic Growth, 1975–2000," in *Bell Journal of Economics and Management Science*, Vol. 5 (Autumn 1974), pp. 461–514; see also Federal Energy Administration, *Revised Base Case Forecast and the President's Program Forecast*, Technical Report 75-2 (5 February 1975), Table A-1, p. 33.

49. OECD, *Energy Prospects to 1985*, Vol. 2, Table 2C-1, p. 11. The table indicates a price elasticity of −0.25 and −0.23 for the United States and Western Europe, respectively, over an eight-year period, 1972–80. For the present analysis, which covers the four-year period 1974–77, the OECD elasticity was cut in half, hence the price elasticity of −0.12.

50. This assumption means that if energy prices rose 10 percent, consumption of energy would fall 0.3 percent a year, bringing about a 1.2 percent decrease in consumption after four years.

51. The demand for energy could, of course, be affected by governmental conservation measures (such as a tax on gasoline), as well as by consumer reactions to high prices. The projections presented in this section take no account of official measures, except to the extent that governmental conservation policies in 1974 may have reduced energy consumption during that year. These projections also do not take into account those increases in taxes on refined products that went into effect in 1975.

There follows an analysis of possible trends in energy consumption for the period 1975–77, made on the basis of the GNP growth rates and the GNP and price elasticities described above, in the three major industrialized areas. It is assumed that the average 1974 prices of refined products (including taxes) which prevailed in each region will be maintained in real terms through 1977. Changes in the prices of refined petroleum products, for which adequate data are available, are used as proxies for changes in the prices of all forms of energy. Their use is justified by the fact that most other energy prices have in fact moved upward with only a slight lag to levels commensurate with the prices of petroleum products.

Consumption of coal, natural gas, hydroelectricity, and nuclear power were estimated independently, principally on the basis of projected availabilities of supply, including imports. Oil was assigned the role of residual energy source on the assumption that a continued lag of other energy prices behind the price of refined petroleum products, supplemented by government efforts to encourage the use of substitutes for oil, would create a preference for other sources of energy than oil.

It must be emphasized that the following forecasts of energy consumption in various parts of the world rest on a number of assumptions, some of them quite arbitrary. These forecasts must therefore be regarded as illustrative, rather than predictive.

The United States

The United States is the only area in which a negative rate of growth in GNP is expected for two consecutive years. The pattern which emerges from the base-case forecast in Table 7-5 is one of a decline from 1973 levels in total consumption of both energy and oil in 1974 and further declines in consumption of both energy and oil in 1975, with a rise in consumption in 1976 as the economy starts to improve, and further increases in consumption in 1977. According to this forecast, energy consumption in 1976 will still be below 1973 levels. In 1977, the second year of recovery, both oil consumption and total energy consumption will be at about 1973 levels in even the low case.

It should be noted that for 1977 energy consumption in each of the three cases was based on 1976 consumption in the same case. (That is, 1977 consumption in Case I is based on the 1976 consumption figure in Case I.) Other sequences are of course possible, but for the sake of simplicity,

Table 7-5. *Energy Consumption in the United States in 1973 and 1974, with Forecasts for 1975, 1976, and 1977*

Consumption in millions of barrels a day of oil equivalent; GNP growth rate in percent

Year and component	Case I	Base case	Case II
1973			
Energy consumption	. . .	36.8	. . .
Oil	. . .	17.3	. . .
Coal	. . .	6.3	. . .
Natural gas	. . .	11.4	. . .
Hydropower	. . .	1.4	. . .
Nuclear power	. . .	0.4	. . .
1974			
Energy consumption	. . .	35.4	. . .
Oil	. . .	16.6	. . .
Coal	. . .	6.2	. . .
Natural gas	. . .	10.7	. . .
Hydropower	. . .	1.4	. . .
Nuclear power	. . .	0.5	. . .
1975			
GNP growth rate	. . .	−3.5	. . .
Energy consumption	. . .	34.9	. . .
Oil	. . .	16.0	. . .
Coal	. . .	6.3	. . .
Natural gas	. . .	10.3	. . .
Hydropower	. . .	1.4	. . .
Nuclear power	. . .	0.9	. . .
1976			
GNP growth rate	4.5	6.5	8.5
Energy consumption	35.8	36.1	36.5
Oil	16.6	16.9	17.3
Coal	6.7	6.7	6.7
Natural gas	9.9	9.9	9.9
Hydropower	1.4	1.4	1.4
Nuclear power	1.2	1.2	1.2
1977			
GNP growth rate	5.5	7.5	9.5
Energy consumption	36.9	37.6	38.3
Oil	17.3	18.0	18.7
Coal	7.1	7.1	7.1
Natural gas	9.6	9.6	9.6
Hydropower	1.4	1.4	1.4
Nuclear power	1.5	1.5	1.5

the computations of oil consumption were not made on the basis of all possible paths of economic growth.

Table 7-5 shows only modest increases in consumption of nuclear power, on account of the various technical, environmental, and other problems that have severely retarded the growth of nuclear power and the stagnant state of demand and the economic problems with which utilities have been confronted. Consumption of coal shows some increase during the period, but not so much as potential output would permit. More dramatic increases in consumption of this fuel are not indicated in the short run because of uncertainties about environmental standards, the difficulties of procuring low-sulphur coal, and overall uncertainty about national energy policy. The increases in coal consumption shown in the table are based on the assumptions that the stricter emissions standards which had been scheduled to be implemented in 1975 will be postponed and

Sources for Table 7-5:

Oil (includes natural gas liquids and bunkers and excludes net changes in primary stocks)

1973 and 1974: U.S. Department of the Interior, Bureau of Mines, "Crude Petroleum, Petroleum Products, and Natural Gas Liquids," *Mineral Industry Surveys*, various issues.

1975, 1976, and 1977: See text for methodology of forecasting total consumption of energy and total consumption of oil.

Coal

1973 and 1974: U.S. Bureau of Mines.

1975: Preliminary estimate of U.S. Bureau of Mines.

1976: Estimate based on the assumption that new air quality standards scheduled to go into effect July 1975 are delayed and that some (but not all) conversion of oil to coal recommended by the Federal Power Commission will go into effect in 1976. The 1976 consumption estimate includes 10 million tons more than the normal 5.5 percent increase in consumption.

1977: Estimate based on the average annual increase in U.S. coal consumption from 1968 through 1972.

Natural Gas

1973: U.S. Bureau of Mines.

1974–75: Federal Power Commission.

1976–77: Domestic production from Federal Power Commission, *A Realistic View of U.S. Natural Gas Supply* (December 1974), pp. 10–13, and from Federal Energy Agency, *Project Independence Report* (November 1974), pp. 93 and 94. Imports from Federal Power Commission, *National Gas Survey*, Vol. I, pp. 45–74. (The price assumption for domestic production of nonassociated gas was 60 cents per thousand cubic feet at the wellhead. According to FEA estimates in the "business-as-usual" case, however, there was little difference in the response of supply to demand at an assumed price of $2.00 or more per thousand cubic feet. The assumed crude oil price for associated natural gas was $11.00 a barrel. See *Project Independence Report*, Tables II-12 and II-13, "business-as-usual" case, pp. 93 and 94.)

Nuclear power

1973: Federal Power Commission.

1974–77: Informal staff estimates, Office of Planning and Analysis, Energy Research and Development Administration (April 1975).

Hydropower

1973: Federal Power Commission.

1974: Edison Electricity Institute, New York.

1975–77: Estimates made on assumption that output of hydroelectric power is not expanding.

GNP growth rates are from Table 7-4. Case I GNP growth rates for 1976–77 are 2 percentage points lower than those in the base case; Case II growth rates are 2 percentage points higher.

The long-term GNP elasticity of demand for energy is assumed to be 1.0 and the long-term trend in GNP growth 4.0 percent annually. The coefficient relating short-term deviations in GNP growth from trend and energy consumption is assumed to be 0.5.

The price elasticity is assumed to be -0.12, applied in equal annual installments of -0.03 for the period 1974–77. The weighted average of petroleum product prices was 54 percent higher in 1974 than in the first nine months of 1973.

In converting kilowatt hours of electric power into oil equivalent, it was assumed that 1 million tons of oil produces about 4 billion kilowatt hours of electricity.

See text for further discussion of methodology.

that some of the utilities capable of burning either coal or oil will switch back to coal, as they did during the emergency created by the Arab oil embargo of 1973–74. Consumption of natural gas is seen as declining over the period, because of diminishing supplies in the forty-eight contiguous states. (The means of shipping gas from Alaska will not be available during this period.)

In sum, the United States appears to have only a small capability for switching to other fuels than oil in the next few years. Patterns of consumption of energy and oil are seen as departing from past trends of rapid annual increases, because of the recession and the large 1973–74 price increases. In the recovery years, 1976 and 1977, consumption of energy in the base case grows at a lower rate (less than 4 percent a year) than GNP because of the assumed relationship between energy and economic activity in an "off-trend" period and because of continuing price effects. Oil consumption grows at a much higher rate than energy consumption, however—about 6 percent a year—because of the absolute declines projected in consumption of natural gas and the very low increases in consumption of energy from coal and nuclear power.

OECD Europe

The base-case forecast in Table 7-6 suggests that there may be little growth in total consumption of energy in Western Europe during the next few years, a reflection of both sluggish-to-moderate economic performance and the impact of the price increase. Oil consumption in the base case levels off in 1976–77 after having declined in 1974 and 1975; oil consumption shows a moderate decline in 1976–77 in Case I and a small increase in Case II. The failure of oil consumption to grow very much even in the high case is attributable to the gradual increase in consumption of energy from other sources—particularly natural gas (with North Sea production rising during the period) and nuclear power. The consumption of coal is projected to rise moderately, in contrast to the continued decrease that had been anticipated before the Arab oil embargo and price increases of 1973–74. The governments of the United Kingdom and West Germany have launched programs to reverse the decline in output from their coal industries, and coal imports from Eastern Europe are expected to increase.

The impact of the substitution of other fuels is not dramatic, however.

In the base-case estimates shown in Table 7-6 oil consumption as a percentage of total energy consumption declines moderately—from 63 percent in 1973 to about 55 percent in 1977. The main effect of the growth in supplies of energy from sources other than oil is that increases in total consumption of energy no longer cause a commensurate increase in demand for oil. If total energy consumption should be greater in 1976 and 1977 than is indicated in the base case, however—and if supplies of energy from sources other than oil are no greater than those shown in Table 7-6 —virtually all the incremental increase would be reflected in rising oil consumption.

Japan

The base-case forecast in Table 7-7 suggests that the level of energy consumption in Japan will have been essentially stable from 1973 through 1975. Despite the economic downturn in 1974, consumption of energy in Japan did not decline, but remained at or very close to 1973 levels. Even so, the recession and the Arab oil embargo together appear to have had an enormous effect on energy consumption in general and oil consumption in particular, in that energy consumption had increased by roughly 11.8 percent annually for the preceding ten years.

Total consumption of energy is expected to increase in 1975, 1976, and 1977, reflecting an upturn in economic activity. Oil consumption continues to decline in 1976, however, owing both to the continued effects of the price increase on total consumption of energy, and to increases in consumption of energy from sources other than oil. In 1977 consumption of oil is projected to rise slightly, but does not attain 1973 levels. The rise in consumption of other fuels is made possible by increases in imports of coal and liquefied natural gas and by a small growth in the use of nuclear power. If economic growth in 1976 and 1977 should be greater than the base-case forecast in Table 7-7, and if total consumption of energy should be greater, most of the incremental increases in consumption would be supplied by increased imports of oil. (Imports of coal could also conceivably be somewhat larger than those indicated in the table.)

Energy Consumption for Other GNP Growth Rates

Tables 7-5, 7-6, and 7-7 provide a range of estimates of energy and oil consumption for 1976–77 for the three major industrialized regions. The

Table 7-6. *Energy Consumption in OECD Europe in 1973 and 1974,*
with Forecasts for 1975, 1976, and 1977

Consumption in millions of barrels a day of oil equivalent; GNP growth rate in percent

Year and component	Case I	Base case	Case II
1973			
Energy consumption	...	23.5	...
Oil	...	14.8	...
Coal	...	5.2	...
Natural gas	...	2.6	...
Hydro and nuclear power	...	0.9	...
1974			
Energy consumption	...	23.3	...
Oil	...	14.1	...
Coal	...	5.1	...
Natural gas	...	3.1	...
Hydro and nuclear power	...	1.0	...
1975			
GNP growth rate	...	2.2	...
Energy consumption	...	23.4	...
Oil	...	13.7	...
Coal	...	5.3	...
Natural gas	...	3.3	...
Hydro and nuclear power	...	1.1	...
1976			
GNP growth rate	2.2	4.2	6.2
Energy consumption	23.5	23.8	24.1
Oil	13.2	13.5	13.8
Coal	5.3	5.3	5.3
Natural gas	3.7	3.7	3.7
Hydro and nuclear power	1.3	1.3	1.3
1977			
GNP growth rate	3.4	5.4	7.4
Energy consumption	23.8	24.4	25.0
Oil	12.9	13.5	14.1
Coal	5.3	5.3	5.3
Natural gas	4.0	4.0	4.0
Hydro and nuclear power	1.6	1.6	1.6

Sources: 1973 data, except for hydro and nuclear power, are from *Statistics of Energy 1959–1973* (Paris: Organisation for Economic Co-operation and Development, 1974).

The figures for hydro and nuclear power are derived from a European Community (EC) working paper, *The Energy Situation in the Community: Situation 1974, Outlook 1975* (XVII/65/75; Brussels, 1975); it was assumed that the EC share of Western European consumption of hydro and nuclear power in 1973 was 90 percent.

Energy consumption for 1974 was also derived from the EC working paper cited above, which shows a range was derived from a base-case forecast of economic activity, with alternative forecasts based on assumptions of GNP growth rates 2 percentage points higher and lower than the base case. The economic outlook could change in any of the three regions, however, thereby giving occasion for revised forecasts not necessarily falling within the assumed ranges. It may be useful to show the relationship between percentage changes in GNP and quantities of oil and energy consumption implicit in the alternative forecasts in the preceding regional tables.

A change of 1 percentage point in the real growth rates of GNP in the three major industrialized areas during the period 1976–77 may be associated with the following changes in the consumption of both energy and oil,[52] in thousands of barrels a day of oil or the equivalent:

United States	260
OECD Europe	225
Japan	100

The differing results for the three areas reflect differences in the size and energy intensiveness of their economies and—of particular importance at present—differences in the assumed responsiveness of energy consumption to short-term fluctuations in GNP. The estimated changes in consumption of energy and oil shown above are applicable only for periods of recession or recovery from recession.

52. The figures for energy and oil are identical, because oil has been treated as the residual source of energy, and energy from other sources has been held constant.

decline of 0.67 percent from the 1973 level; it was assumed that consumption of energy in all of OECD Europe fell by the same percentage.

The method used in forecasting total consumption of energy and oil in 1975, 1976, and 1977 is explained in the text.

Estimates of consumption of energy from sources other than oil for 1974 and 1975 were derived from the EC working paper cited above by assuming that the EC share of consumption of coal, natural gas, and hydro and nuclear power were 84 percent, 92 percent, and 90 percent, respectively, of Western European consumption. Consumption of coal (including lignite), after declining in 1974 (primarily because of the British coal strike), is projected to return to the 1973 level in 1975.

In the absence of clear evidence that consumption of coal will either decline or rise, consumption was assumed to remain stable in 1976 and 1977. Consumption of natural gas for 1976 and 1977 was computed on the assumption that consumption would grow at a rate of about 10 percent a year; for hydro and nuclear power the assumed annual growth rate was about 20 percent a year.

GNP growth rates are from Table 7-4. Case I GNP growth rates for 1976–77 are 2 percentage points lower than those in the base case; Case II growth rates are 2 percentage points higher.

The long-term GNP elasticity of demand for energy is assumed to be 1.0 and the long-term trend in GNP growth 4.5 percent annually. The coefficient relating short-term deviations in GNP growth from trend and energy consumption is assumed to be 0.7.

The price elasticity is assumed to be −0.12, applied in equal annual increments of −0.03 for the four-year period 1974–77. The weighted average of petroleum product prices was 83 percent higher in 1974 than in the first nine months of 1973.

In converting kilowatt hours of electric power into oil equivalent, it was assumed that 1 million tons of oil produces about 4 billion kilowatt hours of electricity.

Table 7-7. *Energy Consumption in Japan in 1973 and 1974,*
with Forecasts for 1975, 1976, and 1977

Consumption in millions of barrels a day of oil equivalent; GNP growth rate in percent

Year and component	Case I	Base case	Case II
1973			
Energy consumption	. . .	6.9	. . .
Oil	. . .	5.3	. . .
Coal	. . .	1.1	. . .
Natural gas	. . .	0.1	. . .
Hydropower	. . .	0.4	. . .
Nuclear power	. . .	neg.	. . .
1974			
Energy consumption	. . .	6.8	. . .
Oil	. . .	5.0	. . .
Coal	. . .	1.2	. . .
Natural gas	. . .	0.1	. . .
Hydropower	. . .	0.4	. . .
Nuclear power	. . .	0.1	. . .
1975			
GNP growth rate	. . .	4.1	. . .
Energy consumption	. . .	7.0	. . .
Oil	. . .	4.8	. . .
Coal	. . .	1.3	. . .
Natural gas	. . .	0.2	. . .
Hydropower	. . .	0.4	. . .
Nuclear power	. . .	0.3	. . .
1976			
GNP growth rate	3.3	5.3	7.3
Energy consumption	7.1	7.2	7.4
Oil	4.6	4.7	4.9
Coal	1.4	1.4	1.4
Natural gas	0.3	0.3	0.3
Hydropower	0.4	0.4	0.4
Nuclear power	0.4	0.4	0.4
1977			
GNP growth rate	5.5	7.5	9.5
Energy consumption	7.4	7.6	7.9
Oil	4.7	4.9	5.2
Coal	1.5	1.5	1.5
Natural gas	0.4	0.4	0.4
Hydropower	0.4	0.4	0.4
Nuclear power	0.4	0.4	0.4

Alternative Assumptions concerning Price Elasticity

Because little is known about the short-run effects on demand for energy of a quadrupling of prices, alternative projections of demand were made on the assumption that the price elasticity was 100 percent greater than that used in the base-case forecasts. The total price elasticity over the four-year period (1974 through 1977) was assumed to be -0.24, spread evenly over the four years—that is, the annual incremental price elasticity was -0.06. Substitution of this higher elasticity resulted in a net reduction in the total demand for energy in the three major industrialized regions by 1977 of the equivalent of about 4 million barrels of oil a day from the base-case forecast for 1977. Oil consumption declined by the same amount, since consumption of energy from sources other than oil is the same in all forecasts, with oil assigned the role of residual supplier (see Table 7-8).

Alternative assumptions about energy demand/GNP elasticities that would result in a different set of total energy forecasts could also be made. And total oil consumption will be altered if consumption of energy from sources other than oil follows a different path from that shown in Tables 7-5, 7-6, and 7-7. The biggest question is probably that concerning nuclear power, production of which has for many years been expected to surge ahead at almost any time.

Not much is likely to be gained, however, from the development of a large number of forecasts by changing assumptions concerning each of the variables and applying the changes in different combinations. The

---

Sources for Table 7-7:

1973: Oil consumption is from *Statistics of Energy 1959–1973* (Paris: Organisation for Economic Co-operation and Development, 1974), pp. 28–29, and from Statistical Office, Department of Economics and Social Affairs, United Nations. Consumption of energy from all other sources supplied by the Institute of Energy Economics (Japan).

1974: Oil consumption figure is based on data on inland consumption supplied by the Institute of Energy Economics plus independent estimates of consumption in bunkers and of crude oil burned in electric power plants. Estimates of consumption of energy from all other sources supplied by the Institute of Energy Economics.

1975–77: See text for methodology for forecasting total energy and oil consumption. All estimates for consumption of energy from sources other than oil supplied by the Institute of Energy Economics.

GNP growth rates are from Table 7-4. Case I GNP growth rates for 1976–77 are 2 percentage points lower than those in the base case; Case II growth rates are 2 percentage points higher.

The GNP elasticity of demand for energy is assumed to be 1.0 and the long-term trend in GNP growth 6.0 percent annually. The coefficient relating short-term deviations in GNP growth from trend and energy consumption is assumed to be 0.9.

The price elasticity is assumed to be -0.12, applied in equal annual increments of -0.03 for the four-year period 1974–77. The weighted average of petroleum product prices was 63 percent higher in 1974 than in the first nine months of 1973.

In converting kilowatt hours of electric power into oil equivalent, it was assumed that 1 million tons of oil produces about 4 billion kilowatt hours of electricity.

Table 7-8. *Base-Case Forecast of Energy and Oil Consumption Compared with Forecast Using a Doubling of the Assumed Price Elasticity*[a]

Consumption in millions of barrels a day of oil equivalent; GNP growth rate in percent

Country or region and variable	1973	1974	1975[b]		1976[b]		1977[b]	
			A	B	A	B	A	B
United States								
GNP growth rate	5.9	−2.2	−3.5	−3.5	6.5	6.5	7.5	7.5
Energy	36.8	35.4	34.4	34.9	35.0	36.1	35.8	37.6
Oil	17.3	16.6	15.5	16.0	15.8	16.9	16.2	18.0
Western Europe								
GNP growth rate	5.4	2.8	2.2	2.2	4.2	4.2	5.4	5.4
Energy	23.5	23.3	22.8	23.4	22.6	23.8	22.6	24.4
Oil	14.8	14.1	13.1	13.7	12.3	13.5	11.7	13.5
Japan								
GNP growth rate	10.2	−3.3	4.1	4.1	5.3	5.3	7.5	7.5
Energy	6.9	6.8	6.8	7.0	6.9	7.2	7.1	7.6
Oil	5.3	5.0	4.6	4.8	4.4	4.7	4.4	4.9
Total energy consumption	67.2	65.5	64.0	65.3	64.5	67.1	65.5	69.6
Total oil consumption	37.4	35.7	33.2	34.5	32.5	35.1	32.3	36.4

Sources: Based on Tables 7-5, 7-6, and 7-7.

a. Figures for 1973 and 1974 show actual consumption and GNP growth rates for these years. Forecast for 1975–77 is shown for base-case GNP growth rates only (see Table 7-4). The alternative price elasticity used is the only variable which is changed from the regional base-case forecasts shown in Tables 7-5, 7-6, and 7-7.

b. Figures in Column A are based on the assumption that the incremental price elasticity is −0.06 per year; figures in Column B are the base-case forecast, where the price elasticity is −0.03 per year.

most important variables affecting the demand for energy in the short run are the level of economic activity and the price elasticity of demand. Alternative assumptions to the base case were made for the purpose of presenting a range of possible demand for energy in the major industrialized areas. Conservative assumptions about the consumption of energy from sources other than oil were made (particularly with respect to nuclear power), in order to show the most difficult situation with respect to oil imports that the major consuming nations are likely to face over the next two or three years.

Other Oil-Importing Countries

In both 1973 and 1974, the United States, Western Europe, and Japan accounted for 85 percent of the oil consumed by the non-Communist nations that are net importers of oil. Most of the remainder was consumed by developing nations. The other oil-importing industrial countries—Australia, New Zealand, and South Africa—consumed only about 0.9 million barrels of oil a day in 1973 and 1974, a little more than 2 percent of total consumption in the oil-importing non-Communist countries.

The net oil-importing developing countries as a group were extremely vulnerable to the fourfold increase in world crude oil prices in 1973 and 1974 because of the impact on their respective balances of payments, which in a number of instances were already under severe strain before the rise in the price of oil. Oil consumption in the developing countries apparently did not decline in 1974, primarily because many of the more advanced developing countries continued to benefit from the worldwide commodity boom that continued through much of the year. The worldwide recession is expected to prevent any increase in oil consumption by the developing countries in 1975.

In 1976 and 1977 oil consumption and imports are expected to rise, with the developing countries following the industrial countries in a general economic recovery. While many of the poorest developing countries (those with per capita incomes of less than $200—such as India, Bangladesh, and most of the countries in Africa and Central America) are likely to experience little real economic growth in 1976 and 1977, consumption of oil by this group accounts for a relatively small share of total consumption of oil in the developing countries. Furthermore, because most of the energy consumption in the developing countries is non-discretionary and the opportunities of these countries for switching from oil to other sources of energy in the short run are limited, it is assumed that, by one means or another, oil will continue to be imported by this group, although at a lower rate of increase than before 1973. It is assumed that economic growth will revive in those developing countries whose economic growth is linked to the capacity to export to the industrialized world and whose consumption of oil accounts for a considerable portion of the total in the developing world.

The other industrial countries appear to have been relatively little affected by the current recession. Australia's GNP grew 3.5 percent in 1974, and strong signs of a rise in economic activity appeared early in 1975. In 1974, South Africa recorded a 7.2 percent increase in real GNP, well above the average rate of 5.75 percent called for in the government's plan for 1972–77.[53]

53. Information on the economies of Australia and South Africa was obtained from the embassies of those countries in Washington.

Table 7-9. *World Oil Consumption and Import Requirements in 1973 and 1974, with Forecasts for 1975, 1976, and 1977*[a]

Millions of barrels a day

Country or region and variable	1973	1974	1975 Base case	1976 Case I	1976 Base case	1976 Case II	1977 Case I	1977 Base case	1977 Case II
United States									
Consumption	17.3	16.6	16.0	16.6	16.9	17.3	17.3	18.0	18.7
Production	10.9	10.5	10.2	10.1	10.1	10.1	10.0	10.0	10.0
Import requirements	6.4	6.1	5.8	6.5	6.8	7.2	7.3	8.0	8.7
OECD Europe									
Consumption	14.8	14.1	13.7	13.2	13.5	13.8	12.9	13.5	14.1
Production	0.4	0.5	0.9	1.2	1.2	1.2	1.8	1.8	1.8
Import requirements	14.4	13.6	12.8	12.0	12.3	12.6	11.1	11.7	12.3
Japan									
Consumption	5.3	5.0	4.8	4.6	4.7	4.9	4.7	4.9	5.2
Production	...	...	...	...	...	...	...	...	...
Import requirements	5.3	5.0	4.8	4.6	4.7	4.9	4.7	4.9	5.2
Other industrial countries									
Consumption	0.9	0.9	1.0	0.9	1.0	1.1	1.0	1.1	1.2
Production	0.4	0.4	0.4	0.4	0.4	0.4	0.4	0.4	0.4
Import requirements	0.5	0.5	0.6	0.5	0.6	0.7	0.6	0.7	0.8
Developing countries[b]									
Consumption	5.6	5.6	5.6	5.8	6.0	6.2	6.3	6.5	6.7
Production	1.5	1.6	1.7	1.9	1.9	1.9	2.1	2.1	2.1
Import requirements	4.1	4.0	3.9	3.9	4.1	4.3	4.2	4.4	4.6
Total, all areas									
Consumption	43.9	42.2	41.1	41.1	42.1	43.3	42.2	44.0	45.9
Production	13.2	13.0	12.9	13.6	13.6	13.6	14.1	14.1	14.1
Import requirements	30.7	29.2	27.9	27.5	28.5	29.7	27.9	29.7	31.6

Sources:

Major industrial countries

Figures for oil consumption in the United States, OECD Europe, and Japan are based on Tables 7-5, 7-6, and 7-7, respectively.

Figures for domestic oil production in the United States are from the following sources: 1973—U.S. Bureau of Mines, *Mineral Industry Survey*, various issues; 1974—data provided the authors by the U.S. Bureau of Mines; 1975—*Oil and Gas Journal*, 27 January 1975, p. 106; 1976—industry sources; 1977— Federal Energy Administration, *Project Independence Report*, November 1974, Table II-8, p. 81 (business-as-usual case, $11 a barrel). Production in 1977 includes 100,000 barrels a day of production from the North Slope of Alaska which may not materialize until 1978.

In the case of OECD Europe, it was assumed the 0.4 million barrels a day produced in 1973 would be maintained through 1977 and that all increases above that level would come from the U.K. and Norwegian sectors of the North Sea. Production figures for OECD Europe are from the following sources: 1973— *Statistics of Energy, 1959–1973* (Paris: Organisation for Economic Co-operation and Development, 1974), pp. 20–21; 1974—authors' estimates; 1975, 1976, and 1977—*Petroleum Economist*, Vol. 42 (February 1975), pp. 53–54, for the U.K. sector of the North Sea, and Embassy of Norway, Washington, D.C., for Norwegian sector.

Other industrial countries include Australia, New Zealand, and South Africa. The 1973 estimates of oil consumption are from the United Nations, Department of Economic and Social Affairs, Statistical Office. It was assumed that during the period 1974–77 oil consumption will grow at an annual rate of 5 percent.

Only Australia is an oil producer among this group of countries. It was assumed that Australia would continue to produce about 400,000 barrels of oil a day during the period 1974–77. The figure for Australian production in 1973 is from the OECD report cited above, pp. 30–31.

For this group of countries oil consumption in Case I was arbitrarily set 100,000 barrels a day lower than consumption in the base case, and in Case II it was set 100,000 barrels a day higher.

Developing countries

The base-case estimates of oil consumption and production in developing countries that are net oil-importers were derived from long-term projections by Edward R. Fried in Joseph A. Yager, Eleanor B. Steinberg, and others, *Energy and U.S. Foreign Policy* (Ballinger Publishing Company, 1974). (See especially Table 13-7, p. 252, and Table 13-8, p. 257.) The levels of oil consumption implicit in Fried's projections have been adjusted downward to take account of the worldwide recession. Oil consumption figures include the following amounts for bunkers (in millions of barrels a day): 1973—0.7; 1974—0.7; 1975—0.7; 1976—0.8; 1977—0.9.

a. Figures for oil consumption include bunkers. Import requirements for all areas are the difference between estimated consumption and domestic production; because of changes in stocks, they do not necessarily equal imports. Net oil exporters and Communist countries are not included in this table.

b. For the developing countries, consumption in Case I was arbitrarily set 200,000 barrels a day lower than consumption in the base case; in Case II it was set 200,000 barrels a day higher.

The base-case forecast assumes that in the period 1974–77 the GNP of the countries in this group will grow at a fairly steady annual rate of 5 percent. This rate is slightly below the average annual growth rates of 5.3 and 5.4 percent recorded by Australia and South Africa, respectively, in the 1960s. The base-case forecast also assumes that consumption of energy will grow at the same rate as GNP and that oil will provide a constant proportion of total energy. Oil production (which is provided entirely by Australia) is assumed to remain at roughly the present level. As a consequence, oil imports for this group of countries are projected to rise moderately to about 0.7 million barrels a day by 1977.

Global Prospects

In Table 7-9 the results of the above analysis of the demand for oil in various parts of the non-Communist world are summarized, and estimates of domestic production and import requirements are presented. A range of possible outcomes for 1976 and 1977 is arrived at on the basis of several different assumptions concerning the level of economic activity. It must be emphasized, however, that an even wider range of outcomes is conceivable, particularly if the levels of economic activity in the industrial countries diverge even more widely from the base case than has been assumed in the two alternative cases presented in Table 7-9.

It is suggested in Table 7-9 that, given the assumptions stated, the decline in the size of the international oil market that began in 1974 will probably continue in 1975. Even with an upturn in economic activity in 1976, the market may not regain the 1973 level until 1977. Prospects for the import sector of the market—which is of course of greatest interest to OPEC and its members—are much the same as for the market as a whole. This is so because (as indicated in Table 7-9) domestic oil production in the oil-importing nations is expected to change very little in the period 1973–77.

The problem faced by OPEC during these years may in fact be somewhat worse than is suggested in Table 7-9. The members of OPEC at present supply over 90 percent of the import market. The balance is provided by Canada, the Soviet Union, and a number of small exporters, including China. There is a real possibility that in future years OPEC's share of the market will be nibbled away to a small, but still significant, extent.

Possible Future Developments

The governments of the oil-exporting countries and the staff of OPEC in Vienna undoubtedly have also been trying to forecast future developments in the international oil market. OPEC and its members have seen that, at present prices in real terms, the market for their oil in 1975 continued to shrink, as it did in 1974. They may see reason to hope for some growth in the oil market in 1976 and 1977, but they must be uncertain concerning both the timing and the magnitude of such growth.

If OPEC were a monolithic entity with the single-minded objective of maximizing its revenues over the long run, it would be difficult enough to predict what its market strategy would be in the uncertain situation now confronting it. Since OPEC is in fact a loosely aligned group of thirteen sovereign nations varying from one another in their political objectives and their economic circumstances, the problem is even more complicated.

The hypothesis to be used here is that over the long run the oil-exporting countries may well be drawn toward price and production policies that will tend to maximize their revenues as a group, but that in the short run those policies may depart significantly from the revenue-maximizing ideal. This is so because their policies at any particular moment are necessarily the product of interacting and sometimes divergent national policies.

The Problems Facing OPEC

Before considering the various market strategies open to OPEC and its members, it will be useful to examine briefly the way that the problems of the oil-exporting countries may look to them.

It is important first of all to realize that during the brief period since the outbreak of Arab-Israeli hostilities in October 1973 the leaders of these countries have been on a psychological roller coaster. Seizing the opportunity provided by the war and the Arab supply restrictions, the oil-exporting countries pushed oil prices to levels previously unthought of. For a time, their market power seemed to be irresistible. By mid 1974, however, production of oil was outrunning consumption, storage capacity was rapidly being filled, and predictions of a glut of oil began to circulate.

The appearance so soon of the threat of a situation of excess supply must have badly shaken the members of OPEC, but they did nothing collectively to meet the threat. A break in the market was avoided, not by resolute group action to reduce production, but principally by allowing the oil companies to reduce liftings to what their marketing systems could absorb at existing prices. As a result, there were shifts in market shares of various oil-exporting countries that not all of them can have found palatable. By early 1975, a certain amount of tension appeared to exist within OPEC because of the desire of some of its members for larger shares of the market. This tension could well increase if the market continues to shrink and if its recovery is delayed or weak. If it does, the consideration of alternative market strategies will become increasingly urgent.

A variety of forces and circumstances combine to make agreement on any strategy quite difficult. Some oil-exporting countries—most notably Saudi Arabia—have large oil reserves and are currently earning oil revenues considerably in excess of their needs. Such countries will be more concerned than other exporters over the possibility that high oil prices will stimulate the development of energy sources outside the control of OPEC and cause OPEC's market to shrink more rapidly in the 1980s and beyond. Other oil-exporting countries, such as Iran, Nigeria, and Indonesia, have relatively limited reserves and can also spend all of their current oil earnings and more. They therefore have shorter time horizons and will give more weight to the maximizing of revenues in the short run and less to the effects of prices on the future market for oil.

Bridging these differences in economic perspectives among the members of OPEC would be hard enough by itself. The difficulty is compounded by the fact that OPEC must work within a volatile political environment. It is inevitable that OPEC's internal deliberations will be influenced by the state of relations among its members and by their foreign policies in matters not directly related to the marketing of oil. Thus, the complex maneuvers of Arab politics and the shifting rivalries among the states bordering the Persian Gulf must affect the price and production policies of the members of OPEC in various obscure ways that make agreement on a co-ordinated market strategy more difficult.

Decision-making within OPEC is further complicated by the need to take increasing account of the interaction between its policies and those of the oil-importing nations. OPEC also faces the tricky problem of dealing with the major oil-importing countries in what could prove to be a series of conferences. (Despite the failure of the Paris preparatory meeting in April 1975, the idea of a producer-consumer conference is by no means dead.) OPEC must establish reasonable negotiating goals, decide how hard to push for linking oil to other economic questions, and cope with the problem of maintaining a common negotiating position in the face of the differing interests and priorities of its members.

The Policies of Saudi Arabia and Iran

By far the greatest weight, both in the marketplace and probably also in OPEC's own councils, is exercised by Saudi Arabia and Iran. The influence of these two countries rests fundamentally on the fact that they occupy first and second place, respectively, among the world's exporters of oil. They are therefore in a better situation than other exporters are to influence conditions in the international market for oil. Within a considerable range of market conditions, they could stabilize prices by restricting their own production and exports. Saudi Arabia, being the larger of the two exporters and having less need for current revenues, could in fact play this role alone in some situations.[54] Saudi Arabia also possesses a much greater ability to expand production (and therefore to bring prices down, should its leaders believe such a move to be desirable) than does Iran.[55]

Both Saudi Arabia and Iran also possess wide political influence that to some extent reinforces the power that they derive from being the largest exporters of oil. The late King Faisal gained a position of leadership among Arab moderates, partly because of his personal qualities, but even more through judicious political use of part of his kingdom's oil revenues. Whether his successor, King Khaled, will occupy a similar position in Arab councils remains to be seen. Oil money and the personal

54. There is some evidence that Saudi Arabia was moving toward the role of residual supplier in early 1975. Saudi production in February and March was 6.8 and 6.6 million barrels a day respectively, substantially below the average 1974 level of 8.0 million barrels a day. See *Petroleum Intelligence Weekly,* 28 April 1975, p. 11.

55. This is so in both an immediate and a long-term sense. Saudi Arabia's existing production capacity is reported to be 11.2 million barrels a day, more than 4 million barrels a day above recent levels of output. See *Middle East Economic Survey,* 14 February 1975, p. 5. Iran's output in March 1975 was only 0.7 million barrels a day below the 1974 peak of 6.2 million barrels a day. See *Petroleum Intelligence Weekly,* 28 April 1975, p. 11. The disparity in the long-run potentials of the two countries is indicated by the fact that Saudi Arabia has proved reserves of 132 billion barrels, more than twice as large as Iran's reserves of 60 billion barrels. See *Oil and Gas Journal,* 31 December 1973, pp. 86–87.

qualities of the monarch also contribute to Iran's political influence, as do its growing military power and the rapid modernizing of its economy.

Recent relations between Saudi Arabia and Iran have in general been good, although Iran's desire for regional hegemony has created at least a latent rivalry between the two nations, and in the 1960s, each put pressure on the oil companies to take more of its oil at the expense of the other. Also, the fact that Iran is a non-Arab nation and does not share Saudi Arabia's strong feelings on the issue of Palestine is an obstacle to really close relations.

In 1974, Saudia Arabia and Iran took different public positions on the question of how high oil prices should be. Saudi spokesmen urged that prices be reduced somewhat in the interests of the health of the world economy. The Shah and his spokesmen argued that prices should be increased further in order to match the rising costs of imports from the industrialized nations. Whether the Saudis and Iranians have differed so sharply within the private meetings of OPEC is not known, but if their public declarations are taken at face value, it appears that neither got its way entirely during 1974.

The price of Arabian light crude oil, f.o.b. Persian Gulf, rose by about 8 percent during the year in current dollars, roughly half the rise in the general price level and not much more than a third of the increase in the cost of imports to the members of OPEC.[56] Thus, while the price of oil fell somewhat in real terms, the decrease did not approach the cut that the Saudis may have had in mind.[57] On the other hand, the nominal increase in the price of oil was far short of the parity with import prices sought by Iran.

One is tempted to conclude that some kind of stand-off had developed within OPEC between Saudi Arabia and Iran, with neither able to impose its will. The real situation was probably much more complicated. Saudi Arabia probably could have forced prices down by cutting its own price and expanding its own output. The Saudis apparently did not want to break with the other OPEC nations, however, or to increase the risk of becoming isolated within the Arab world. As for Iran, its failure to persuade OPEC to push prices higher may have been due to the opposition

56. *Middle East Economic Survey,* 21 February 1975 supplement, p. 3.

57. Sheikh Ahmed Zaki Yamani, the Saudi oil minister, stated publicly in September 1974 that his government favored a reduction of the posted price by $2 a barrel, or about 17 percent. See *Washington Star-News,* 21 September 1974.

of Saudi Arabia, but the effect of softening market conditions on general attitudes within OPEC may have served as an even more powerful constraint.

In any event, by late 1974 the Saudi commitment to lower prices seemed more muted than it had earlier in the year, and Iranian pronouncements in defense of higher prices had lost some of their past hawkish tone. The central question for the future so far as these two leading oil-exporting countries are concerned is not so much which one will prevail in the private councils of OPEC as whether they will co-operate in regulating oil production, either by supporting prorationing or by accepting the role of residual suppliers.

While the policies of Saudi Arabia and Iran will continue to be of paramount importance, these two countries will by no means be in full control of events. Smaller exporters, such as Algeria, Kuwait, and Venezuela, appear on occasion to have possessed an influence within OPEC out of proportion to their shares of the international oil market. Also, any oil-exporting country can gain extra leverage by simply threatening to break ranks and cut prices. This potential leverage is all the greater in the case of a country with substantial excess productive capacity or reserves large enough to support a rapid increase in capacity.

In this connection, the possible future role of Iraq is a subject of considerable interest. Iraq's proved reserves of oil are nearly as large as those of the United States,[58] and Iraqi officials claim that additional very large deposits have been discovered.[59] After a decade of struggling with the Western oil companies that controlled the Iraq Petroleum Company (IPC) Iraq nationalized IPC in 1972.[60] During the ten years or more of bickering, production increased only slightly. In 1973 Iraq reached a settlement with the former owners, and a new working relationship evolved, under which the government and several Western oil companies launched a major expansion in drilling and production. Production during 1974 averaged less than 2 million barrels a day,[61] but an increase to 6 million barrels a day by the early 1980s may be technically possible.[62]

58. *Oil and Gas Journal,* 31 December 1973, pp. 86–87.

59. *Oil and Gas Journal* newsletter, 31 March 1975.

60. Iraq did not, however, nationalize IPC's wholly owned affiliate, the Basrah Petroleum Company (BPC), in 1972. The Dutch and U.S. interests in BPC were nationalized during the 1973–74 embargo, but not the British and French interests.

61. *Middle East Economic Survey,* 21 February 1975, p. v.

62. *Middle East Economic Survey,* 17 May 1974, p. 7.

The government of Iraq is committed to the somewhat more modest goal of increasing output to 4 million barrels a day by 1980–81.[63] If Iraq should be frustrated in its efforts to gain a larger share of the market, it might conceivably break ranks with OPEC and seek to increase its sales by cutting prices. Iraq's willingness to pursue an independent policy was demonstrated by its refusal in 1973–74 to join other Arab oil-exporting countries in restricting production.

Alternative Market Strategies

A number of market strategies are open to the oil-exporting countries. If any of them is to be effective, it must contain not only a policy on the price of oil, but also a means of adjusting production to the quantity that can be sold at the desired price. Three such means of adjusting production are in theory available:

• Formal prorationing (the enforcement by OPEC of production quotas agreed upon by its members).

• The indirect allocation of market shares by the the oil companies as the combined result of their individual decisions as to where best to obtain the crude oil that their refining and marketing systems can absorb at existing prices. These decisions would presumably be based on commercial considerations and on strategic judgments as to the way in which frictions with individual oil producers could be minimized, so that assured sources of supply would be retained.

• The assumption of the role of residual supplier by one or more financially strong oil-exporting countries.

Combinations of these adjustment mechanisms are also possible, especially over a period of time.

At no time during its fifteen-year history has OPEC been able to ration production. As recently as March 1975, at the OPEC summit meeting in Algiers, Saudi Arabia is reported to have blocked the effort of Algeria, Iraq, and Libya supported by Iran, Nigeria, and Venezuela to get OPEC to adopt a co-ordinated system of regulating production.[64] If the members of OPEC continue to be unable to agree on prorationing, the only alternative clearly available is to continue the strategy of muddling through by placing major reliance upon the oil companies to adjust sup-

63. *Herald Tribune* (Paris), 25–26 January 1975, p. 75.
64. *Middle East Economic Survey,* 7 March 1975, supplement, p. 2.

plies to what the market can absorb. Certainly, OPEC cannot require any of its members to act as residual suppliers. It is entirely possible that a strategy of muddling through could be sustained for several more years, and in the first of the three hypothetical cases examined below it is assumed that OPEC will in fact continue to pursue such a strategy and that it will not be undermined by any serious breaches of market discipline.[65]

In the other two cases, the possible behavior of the oil-exporting countries in two contingencies is considered: a collapse of oil prices brought about by a breakdown in market discipline and the imposition of new production restrictions by the Arab oil-exporting nations in an effort to gain their political objectives with respect to Israel.

Case One: Muddling Through

OPEC's price policy for the greater part of 1975 was set by the announcement, in late 1974, that (nominal) oil prices will remain unchanged until the end of September 1975.[66] Adherence to this price policy as long as the market remained weak would be entirely consistent with the assumed strategy of muddling through.

If nominal oil prices should stay the same through all of 1975 and if the general price level were to rise by 15 percent—which would be higher than is expected—the real price of oil would be 13 percent lower at the end of the year than it was at the beginning. Further small reductions—occasional secret discounts, more generous credit terms, and the ironing out of anomalies such as excessive premiums for low sulphur content—could also be expected and would not necessarily break the market. (All these means of shaving prices had in fact begun to appear in late 1974 and early 1975.[67]) A downward drift in the average price of crude oil (in constant dollars) by as much as 15 percent during 1975 would be quite conceivable. The average price reduction during the year would of course be only about half the total, or 7.5 percent. If it is assumed that all of this reduction would be passed through to final consumers, the real price of refined

---

65. A force working to support market discipline is undoubtedly the fear of the severe criticism that would be levied against price-cutters in the private councils of OPEC.

66. *Middle East Economic Survey,* 7 March 1975, supplement, p. 3.

67. *Wall Street Journal,* 29 January 1975, p. 28; *New York Times,* 5 January 1975; *Oil and Gas Journal* newsletter, 24 February 1975.

products would fall by about 4.4 percent in the United States, 3.4 percent in Western Europe, and 4.5 percent in Japan.

The effect of this fall in product prices on the size of OPEC's market during 1975 would be quite modest. Using even a relatively high price elasticity of −0.1 for the first year after a price cut, the market in the major industrialized countries would be only 0.7 percent, or less than 300,000 barrels a day, larger than it would have been if 1974 prices (in real terms) had continued to prevail. If the base-case forecast presented in Table 7-9 is taken as a rough guide, much larger price reductions could be averted only by reducing the crude oil marketings of OPEC in 1975 by about a million barrels a day below the 1974 level. This would appear to be far from an impossible task.

Whatever distribution of shares of the market emerges from the adjustment to a smaller market in 1974–75, some exporters are certain to be less satisfied than others, and the entire structure will be subject to a certain amount of strain. If the market does not recover in 1976, the strain will of course increase. It might be argued that OPEC will seek to ease the strain by voluntarily cutting prices; this does not appear likely, however. The price hawks in OPEC have been temporarily subdued by declining market conditions, but they have not really been converted. (Even so recently as March 1975 at the OPEC summit conference in Algiers, Iran, with strong support from Algeria, is reported to have fought for the stabilization of oil prices "in real value terms.")[68] Moreover, given the low short-term price elasticity of demand for petroleum products, only a quite substantial price cut would have any great effect on the size of the market.

The willingness of the oil-exporting countries to live with frozen nominal prices—which means declining real prices—for many more months may in fact be seriously questioned. If recovery from the recession in the industrialized countries is accompanied by an immediate upturn in the rate of inflation, some form of indexing of the prices of oil appears probable. If the price of oil were linked to the general price level, indexing would amount to stabilization of the real price of oil. But if, as appears more likely, it were linked to the prices of commodities imported by members of OPEC, indexing could increase the real price of oil by an amount that would depend on which commodities were chosen and how far back in time the index was extended.

When Saudi Arabia and other Persian Gulf producers take over 100

68. *Middle East Economic Survey,* 7 March 1975, supplement, p. 3.

percent ownership of the oil industry within their borders—a development that appears likely in the fairly near future—the distinction between equity and participation crude will disappear. All oil will then presumably be sold at the buy-back price for participation crude, which is somewhat higher than the tax-paid cost of equity crude. At one time it was expected that this move would produce a substantial increase in the price of oil, but the reduction in the posted price (on which the buy-back price is calculated) initiated by Saudi Arabia in November 1974 narrowed the difference. It now appears that 100 percent government ownership will increase the proceeds to governments on the average barrel of oil by only about 3 percent, much less than the prospective rate of inflation.

Assumption of complete ownership of oil company assets by the governments of the countries in which they are located should not materially change the present role of the companies. The governments are already in a position to set production policy, and so long as its members preserve market discipline, OPEC can determine the price of crude oil. (In this sense, 60 percent participation backed by the sovereign power of governments is just as good as 100 percent ownership.) Complete government ownership of company assets will also not affect the ability of the companies to do the prorationing job, if the governments continue to prefer having the companies perform this task.

By 1977, the oil export market should have regained its 1973 size, but prospects for further rapid growth will be clouded by a new factor: the rise in the production of energy that is beyond the control of OPEC. By 1980, 1.5 million barrels a day of Alaskan oil may be flowing into the U.S. market, and North Sea production could reach 4.0 million barrels a day.[69] New exporters, such as China and Mexico, will be increasing their production and pushing their way into OPEC's market. Supplies of energy from other sources—nuclear energy, North American coal and gas, and Soviet gas—are also likely to be increasing.

How the oil-exporting countries will react to this prospect during the next two or three years cannot easily be predicted. It can be argued that even in 1974 it would have been in their best interest over the long term to keep prices low enough not to stimulate the development of other high cost sources of energy. But if they have not adopted such a policy in 1974 and 1975, when the market for their oil has been shrinking, can they be expected to do so in 1977, when it will probably have become easier to

69. *Petroleum Economist,* Vol. 42 (February 1975), p. 44.

maintain high prices? Also, once indexing has been adopted, is there any realistic chance of its being abandoned in favor of a price cut at a time when the market for oil will be improving in a resurgent world economy and the prices of the commodities imported by the members of OPEC will probably be rising?

As market conditions improve, it is in fact likely that an increase, rather than a decrease, in the real price of oil will gain support within OPEC.[70] Whether or not agreement would actually be reached on a significant price increase would depend, however, on the policies of individual oil-exporting countries at the time and on the unpredictable outcome of internal bargaining within OPEC.

A plausible sequence under the assumptions of Case One might therefore be: a modest decline in the real price of oil during 1975, determined largely by the rate of inflation; a price increase of uncertain magnitude in late 1975, possibly in conjunction with indexing; and relative stability through 1977, perhaps depending on the extent to which the oil index deviates from the general price level.

Case Two: Breakdown in Market Discipline

The primary assumption in Case Two is that OPEC's policy of muddling through will end in disaster sometime in late 1975 or 1976. This result could come about in any of a number of ways, but the fundamental cause would in any event be a refusal by one or more oil-exporting countries to accept the share of the market allotted to it as a result of commercial decisions by the oil companies. Those countries would therefore expand production and cut prices to the extent necessary to dispose of the larger volume of oil. Other exporters would lose sales and would be forced to cut prices also as a measure of self-protection. The first price-cutters would find that further cuts were necessary in order to market their increased output, and others would again be forced to follow suit.

Once begun, the process of price-cutting would be hard to stop, especially in a time of excess oil-production capacity.[71] Some oil-exporting countries would be strongly tempted to try to counteract the impact of falling prices on their revenue by drawing on unused capacity and in-

---

70. *Oil and Gas Journal,* 26 May 1975, p. 29.
71. Estimates of the excess in early 1975 ran as high as 10 million barrels a day. See the *Petroleum Economist,* Vol. 42 (March 1975), p. 86.

creasing production. To the extent that they actually did so, the downward pressure on prices would probably increase. Given the low price elasticity of demand for petroleum products, the market could absorb significantly larger offerings only if the market behavior posited here coincided with a strong upsurge of demand in the oil-importing countries.

Such an upsurge is of course exactly what would normally take place as the world began to climb out of the current recession, and this is the point at which the energy conservation policies of the major industrialized countries would become critically important. If oil consumption in those countries during 1976 could be reduced by about one million barrels a day from what it would otherwise have been, there would be little or no growth in the international oil market (as forecast in the base case in Table 7-9) in that year. As a consequence, the oil-exporting countries could not rely on market forces to check the fall in oil prices. They might hope that some of their number would act as residual suppliers, but their willingness to do so would be far from certain given the chaotic market conditions that the assumed breakdown of market discipline would have created. The potential residual suppliers could not be sure that a cut in their own output would not cause others to activate still more idle productive capacity.

The only sure way to stabilize the price of oil would be the institution of a system of rationing production. This might be done by means of an agreement within OPEC alone or as part of a wider understanding between the oil-importing and oil-exporting nations. Neither kind of agreement would be easy to achieve.

If OPEC acted alone to set up a system of prorationing, and if the system worked successfully, an effort would almost certainly be made by some of the members of OPEC to regain the 1974 level of oil prices. Whether it would be possible to do so would depend on the ability of the members of OPEC to agree on the distribution of further cuts in production. Even if this proved to be possible in 1976–77, which is by no means certain, OPEC would be confronted by the need to make still further reductions as supplies of energy not controlled by OPEC increased in the late 1970s.

An agreement between the oil-exporting and oil-importing nations to stabilize the price of oil would be attractive to the exporters as a means of checking cutthroat competition, and it would give the importers some insurance against sudden increases in the price of an essential commodity. The price set would be a matter for negotiation, but would presumably be below the price prevailing in early 1975. Some form of indexing designed to keep the real price of oil stable would probably be part of any agreement between exporters and importers.

Prospects for oil prices under the assumptions of Case Two are therefore even less clear than under those of Case One. The two cases begin in much the same way, with the real price of oil falling at a rate determined largely by the rate of inflation. The breakdown in market discipline assumed in Case Two would be most likely to occur in 1976, but the extent and duration of the consequent fall in oil prices cannot be estimated even approximately. If and when OPEC succeeded in instituting a system of prorationing, the fall in prices would be checked and might even be reversed, although restoration of the 1974 level does not seem likely. If the alternative of an agreement between exporters and importers were successfully pursued, the real price of oil might be stabilized at a negotiated level for several years.

Case Three: Renewed Arab Resort to the Oil Weapon

If major hostilities should again break out between the Arabs and Israelis, a new Arab embargo and supply cutbacks would be a virtual certainty. Even in the absence of major hostilities, the Arabs might turn once more to their oil weapon, if diplomatic efforts to achieve an Arab-Israeli settlement should reach an impasse.

The effects of politically motivated restrictions of oil supplies would depend on their timing, as well as on their depth and duration. Restrictions by the Arabs in 1975, when considerable excess capacity exists,[72] would be less less effective than were those of 1973–74. (By 1976 and 1977, when the oil market will probably again be growing, the impact of supply restrictions could be somewhat greater.)

The emergency system of oil allocation worked out by the IEA is another factor not present in 1973–74. If this system should actually be put into operation in response to new Arab restrictions, the ability of the major oil-importing countries to endure the restrictions for a prolonged

---

72. Although most excess capacity appears to be in Arab hands, significant amounts exist in Iran, in Venezuela, and possibly elsewhere. Iraq, moreover, which did not join the other Arab oil exporters in restricting production in 1973–74, could probably increase output fairly quickly.

period would be enhanced. Also, effective management of the emergency allocation system would in effect tailor demand to available supply and reduce the pressure on the price of oil. Whether some increase in price could be avoided may, however, be doubted. The longer supply restrictions lasted (and the deeper they cut) the more likely a rise in the price of oil would become. When the supply restrictions were lifted, as they eventually would be, the oil-exporting countries would have to decide whether to try to defend the higher prices established while the restrictions were in force.

A second experience with Arab supply restrictions should impel the major oil-importing countries to apply drastic conservation measures and to redouble efforts to develop alternative sources of energy. If they did so, these policies in combination with the normal reaction of consumers to high prices would in time cause the market for OPEC oil to shrink. The oil-exporting countries would then find themselves confronting an even worse set of alternatives than they do today.

Conclusions

To attempt to predict the behavior of the international oil market in the next few years would be foolhardy. The memory of the unanticipated events in the last part of 1973, if nothing else, should inspire both caution and humility. The foregoing analysis does, however, point toward a number of highly tentative conclusions:

Oil prices in constant dollars may continue to drift downward through much of 1975, and possibly longer, at a rate that will be determined largely by the rate of inflation. At some point in late 1975 or 1976, this downward drift may be interrupted in one of two ways:

The end of the recession in the industrialized countries and the consequent revival of the international oil market may encourage OPEC to raise oil prices moderately and possibly also to link the price of oil to an index calculated from the prices of a bundle of the commodities that are imported by the members of OPEC.

Should the real price of oil be increased, the size of the world oil market would be further restricted. This constraint would be even stronger if the recession should be prolonged. In these circumstances dissatisfaction on the part of some oil-exporting countries with their shares of a stagnant market could lead to a breakdown of market discipline and a collapse of oil prices. Stability might be restored if several exporting countries were to act as residual suppliers. It is more likely, however, that the members of OPEC would be forced to agree on a system of prorationing, either on their own initiative or as part of a wider understanding with the oil-importing nations. In the former event, the price of oil might gradually be pushed back toward the 1974 level. In the latter event, stabilization of the real price of oil at some point below the 1974 level would be possible.

If resumption of the Arab-Israeli war brought new Arab supply restrictions, and if those restrictions were both deep and prolonged, some increase in the real price of oil could not be avoided. Whether the oil-exporting countries would try to maintain higher prices after the restrictions had been lifted cannot easily be predicted. The task would not, however, be an easy one, particularly if the major oil-importing nations made strong efforts to conserve energy and expand alternative sources of energy supplies.

Index